Complex Spatial Systems:
The Modelling Foundations
of Urban and Regional Analysis

Complex Spatial Systems: The Modelling Foundations of Urban and Regional Analysis

Alan G Wilson

An imprint of **Pearson Education**

Harlow, England · London · New York · Reading, Massachusetts · San Francisco · Toronto · Don Mills, Ontario · Sydney
Tokyo · Singapore · Hong Kong · Seoul · Taipei · Cape Town · Madrid · Mexico City · Amsterdam · Munich · Paris · Milan

For Sarah

Pearson Education Limited
Edinburgh Gate
Harlow
Essex CM20 2JE
England
and Associated Companies throughout the World.

Visit us on the World Wide Web at
www.pearsoneduc.com

First published 2000

ISBN 0 582 41896 8

British Library Cataloguing-in-Publication Data
A CIP catalogue record for this book can be obtained from the
British Library

Transferred to digital print on demand, 2008

Typeset by 32 in Sabon and News Gothic
Produced by Pearson Education Asia Pte Ltd.,

Printed and bound in Great Britain by CPI Antony Rowe, Eastbourne

Contents

Preface

Cities and regions are of major importance in the contemporary world. They have been the subject of analysis for many decades and huge advances in understanding have been achieved since the 1950s. However, there is no adequately-articulated body of theory – as, say, in physics or chemistry. Urban and regional analysis is essentially interdisciplinary, and this is part of the problem: different perspectives have not been forced together. Cities and regions are immensely complicated; achieving effective theory – which can then be applied in policy-making and planning – has represented, and represents, one of the great scientific challenges of the century. The aim of this book is to present one under-developed set of approaches to urban and regional analysis which have enormous potential for the future. The presentation is set within two broader contexts: first, the challenges of interdisciplinary social science; and secondly, the framework of the rapidly developing field of complexity theory in the sciences more broadly – and hence the title, *Complex Spatial Systems*.

The approaches deployed are based on a particular style of mathematical modelling which has been developed over the last 30 years – models of spatial interaction and associated location models. It is not the intention here to present any of these in detail, but rather to summarise the main ideas as platforms for further development. Most fields of academic study – and perhaps particularly in the social sciences – are subject to the vagaries of fashion. Urban and regional model-based analysis was at the height of fashion in the late 1960s and early 1970s. It then fell out of favour. There has been something of a resurgence in the 1990s as it

has come to be appreciated that such analysis has a major contribution to make to contemporary problems, ranging from traffic congestion to social polarisation and its consequences. It is also the case that developments in science generally, particularly in the areas of complexity theory and computer simulation, help to bring it back into the mainstream. Indeed, it can now be seen that the developments of the 1960s and 1970s can be interpreted as heralding many of these ideas and that urban and regional modelling represent real applications of the ideas of complexity theory.

There is a need, therefore, to introduce the broad ideas to new audiences in all the disciplines which impact on urban and regional studies, to set these ideas in this broader context and, above all, to begin to articulate the rich seams of new research problems that are now visible, waiting to be mined. There is also an argument which is tackled in the book that urban and regional modelling has a major role to play in extending traditional, classical, theories of cities and regions. These theories are therefore articulated in some detail as a basis for rewriting them in much more powerful forms. At the root of this approach is the belief that there are concepts which transcend disciplines – supradisciplinary concepts – which, when understood, extend analytical capability in significant ways. It can be conjectured that this is the case on a broad basis. It is hoped that by presenting one field of study from such a perspective, there are more general lessons to be learned which can be applied in teaching and learning and in research design. As the research programme progresses, the goal of building an integrated and effective theory will be within sight.

The ideas presented in this book have been developed over several decades, many of them, as noted in the relevant references, developed and written up jointly with colleagues in Leeds and elsewhere. I am very grateful to all of them for these collaborations: Bob Bennett, Sergio Bertuglia, Mark Birkin, Arthur Champernowne, Graham Clarke, Martin Clarke, Jose Coelho, Britton Harris, Yi-Xi Jin, Paul Keys, Mike Kirkby, Giorgio Leonardi, Sally Macgill, Roger Mackett, Mark Oulton, Carol Pownall, Philip Rees, Tracy Rihll, Martyn Senior, Huw Williams and all those who have been part of this world at Regional Science Association conferences and elsewhere.

Many others have helped in various ways. I thank the members of my present office in Leeds for their support, particularly Sharon Beckram; and Lois Wright, of the School of Geography in Leeds who drew most of the figures. I am grateful to three referees for their comments, particularly the one who will recognise that he suggested the title! Finally, I would like to thank my wife, Sarah, to whom this book is dedicated. She has provided the support which has made this book possible. I hope she thinks the result is worthwhile!

Alan Wilson
May 1999

Acknowledgements

I would like to thank the owners of the copyrights for permission to base a number of the figures in this book on their originals. They include John Wiley and Sons Limited for Figures 6.1, 6.2, 6.3 and 7.6 © John Wiley and Sons Limited, reproduced with permission; The Institute of British Geographers for Figures 7.1, 7.2, 7.4 and 7.5; Pion Limited, London for Figures 7.3 and A2.22; The MIT Press for Figure A2.36.

While every effort has been made to trace the owners of copyright material, in a few cases this has proved impossible and we take this opportunity to offer our apologies to any copyright holders whose rights we may have unwillingly infringed.

1
Introduction

Cities and regions and 'big science'

Understanding cities and regions represents one of the major scientific challenges of our time. They provide the habitats of the world's populations; and urban and regional analysis should provide the underpinnings of well-being in the same way as the biological and medical sciences underpin health and medicine and the physical and engineering sciences underpin industry. But the science of cities and regions has not, as in these fields, been treated as *'big science'*. None the less, the scientific foundations have been laid in the last three decades and the purpose of this book is to present an overview of what is known, and what can be achieved – albeit from a relatively limited perspective. Of course the social sciences are essentially different from the physical sciences. Human behaviour cannot be understood or predicted in the same way. But many features of cities and regions can be understood, and the knowledge productively applied, and the prime aim of this book is to offer a sketch of what has been achieved to facilitate further development.

The approach adopted has a number of distinctive features. Cities and regions are seen as complex systems and it is argued that a general perspective can be adopted which provides a framework for making effective decisions for building system models. The approach has to be interdisciplinary. This framework shows the weaknesses in some of the traditional approaches to urban and regional analysis and provides a viewpoint for the analysis of classical approaches

to urban and regional model-building. The core of the argument is the presentation of a set of mathematical and computer model-building techniques. These are shown to be precursors of *complexity theory* and this means that insights can be offered about complexity theory in one of its detailed applications. The focus, therefore, is on *complex spatial systems*. Classical theory can be rewritten in a very powerful way, and this is important for a number of disciplines. More importantly, a research agenda can be articulated which shows tremendous potential for the future.

Cities and regions as complex systems

It is self-evident that cities and regions are complex systems, and methods exist for representing the theory of such systems, for example as mathematical and computer models, which represent the foundations for analysis and planning. In this century, tremendous progress has been made in the formulation of these models for cities and regions. The late twentieth century is a particularly exciting time to review what has been achieved because it can now be seen that the developments of the last 30 years can be integrated through the ideas of what is becoming a common approach in many fields of science – complexity theory.

Complexity theory perhaps represents the most important territory in contemporary science. However, presentations are not always complete or clear. It is sometimes misnamed – with labels like 'chaos theory'. Some critics belittle it, but are attacking obscure targets – particular and poor representations. Some proponents oversell it. The danger is that important directions for academic development are inadequately charted. It will be argued that there is tremendous potential for urban and regional analysts which can be unlocked through complexity theory and new interdisciplinary collaboration. Interestingly, it can be argued that this presentation is important to complexity theory as well as to the substantive field for two reasons: first, because in all the excitement of the new developments in complexity theory, the social sciences have been seriously neglected; secondly, because urban modelling in particular demonstrates how the ideas of complexity theory can be made to work in a real context.

We therefore need a good broad understanding of what complexity theory is about. The issues can be usefully addressed in two stages: the substantive subject matter, which is discussed here in terms of *systems*; and the *methods*. There is a preliminary exploration of these topics in the rest of this section. More detailed arguments are presented to complete the preliminaries in Chapters 2–4, and the ideas developed will be used in the context of examples in the rest of the book.

Components of cities and regions can be called *systems* – simply because they involve a large number of interacting components. Complexity theory can then be thought of as theory about complex systems. Urban and regional analysis can then be seen as concerned with complex spatial systems. There are then two initial questions: what is distinctive about *complex* systems? What is distinctive about the *theory* of complex systems?

It was Warren Weaver in the 1940s and 1950s who (to the author's knowledge) (Weaver, 1948, 1958) first introduced a useful distinction between simple and complex systems. In the scientific context, simple systems were those describable by a small number of variables; complex systems needed a large number of variables to describe them. He made a further subdivision of complex systems into those of *disorganised* complexity and those of *organised* complexity. It should now be recognised that a particularly important subset of systems of organised complexity (perhaps the whole set?) are nonlinear systems. Nonlinearities can arise in a variety of ways: when rates of change are anything other than constant; for the geographer, when distance effects, as in the gravity model for instance, involve a power or an exponential function. It is the nonlinearities that are at the basis of what is interesting in complex system behaviour. It turns out that a number of important analytical issues associated with urban and regional systems can be solved using methods for systems of *disorganised* complexity. However, the most interesting problems, as in most other sciences, are those of systems of organised complexity.

What Weaver observed was that problems associated with simple systems could be solved by essentially the mathematics associated with, for example, Newtonian mechanics; the problems of systems of disorganised complexity by the mathematics of statistical mechanics; but, at the time, there were no mathematical solutions to problems of organised complexity. It is the systems of organised complexity, the nonlinear systems, which can, in current parlance, be thought of as *complex* systems – and some of the mathematics does now exist. Weaver, when he was writing in the 1950s, was the Science Vice-President of the Rockefeller Foundation, and he went on to argue that the Foundation, on the basis of this analysis, should be investing more of its funds in biological rather than physical sciences – a prescient analysis! Social scientists, of course, would now wish to be added to Weaver's 'interesting' list!

What characterises systems of organised complexity is essentially that they are made up of large numbers of parts – and that these parts are strongly connected; that is, they each *interact* strongly with a number of others. Obvious examples of systems of organised complexity are human beings, brains, ecosystems, economies and cities. Most of these figure in the popular literature of complexity theory though the social sciences are seriously under-represented. Other dimensions figure in this literature too: time is important (in fields such as evolution); methods, such as nonlinear mathematics and computer science, are important. Many of the ideas employ analogue or metaphor: neural network computing for instance. It should be clear even at this stage of the overall argument that urban and regional systems of interest are typically systems of organised complexity.

What Weaver had not foreseen was the extent to which the methodology to be developed in the four decades since his analysis would be multidisciplinary and generic – and hence the term *complexity theory*. In any particular discipline, it will be particularly important to work towards an understanding of what can be achieved through the deployment of generic tools and what has to be developed which is specific to that discipline.

I will take as a working definition that theories are about understanding systems; and that methods are important elements in theory-building. We should also recognise that most interesting theory-building is concerned with process – the nature of change over time for the system of interest. Scale is particularly important. The same systems can be characterised at different scales, and if we do not insist on absolute clarity in this respect then confusion can ensue. It is obvious intuitively that there are fundamentally different spatial scales at which we can perceive the 'same' systems. It is important to recognise that there are interesting (scientific) phenomena at each of these scales – though sometimes there are important interactions between scales. It can then be argued that the methods which are valuable in theory-building at one scale may be different from those for the same system (or an element of it) at another scale. In the case of temporal scale, as in the study of a biological system for example, the approach will be different if the study is concerned with contemporary function or with evolution over a long time period. In effect, we work with a hierarchy of knowledge about real complex systems. This notion is pursued in more detail in Chapter 2.

KEY IDEA 1.1

Scale is a form of *hierarchy* and clarity of vision in this respect is critical.

This then takes us to *method*. Nonlinearities fundamentally change the nature of the mathematics needed to describe complex systems. This was why Weaver in the 1950s recognised that problems of organised complexity were then insoluble in mathematical terms. What has changed (in the last 20 years or so) is that appropriate mathematical tools have become available. There is now a broad understanding of the mathematics of nonlinear systems and we need to chart out the essence of the ideas involved so that we can understand, at least intuitively, the range of application of each. What is more, as happens in scientific development, these ideas are at least broadly understandable in less technical ways.

Authors such as Holland (1995) argue that there are properties of complex systems which demand a kind of mathematics that is not yet available to us: these properties are based on adaptation. The agents and subsystems which make up cities and regions are capable of adaptation; they *evolve* over time. As we will see later, this capability is very difficult to model.

However, there is another aspect of the methodological tool-kit needed for theory-building which needs to be brought into play here. Most interesting complex systems are very large. So even though some of the mathematics exists in principle, either not enough is known substantively about the system to make mathematical analysis possible, or the system is simply too large for feasible analysis; there are too many variables. This is where another major impact from discoveries of the last 20 years contributes to method: powerful computers. These have meant that many of the problems which are not solvable in *analytical* mathematical terms can be tackled through computer *simulation* – generating great understanding and insight. Much of the power derives from the fact that it is possible to combine human intelligence with computing power – and we should not underestimate the impact of computer graphics, developed with the advent of PC cultures, in this context. In many cases, it is easier to work directly with ideas of computer *modelling* and simulation rather than the more traditional systems of mathematical equations. In the context of urban and regional analysis, these ideas have even manifested themselves in impressive computer games, such as the best-selling SimCity (for a geographer's review, see Macmillan, 1996).

There is a third element to the methodology of complexity theory: the effective deployment of metaphor. This arises from the multidisciplinary power of complexity theory: essentially, in fact, the main concepts are supradisciplinary.

These ideas will allow us to map out (systematically!) in turn the territory of complexity theory and the range of methods which are potentially valuable in systems of interest – in this case, cities and regions.

This argument generates the basis of an approach that will be formalised in the rest of the book, i.e. that there is a three-stage approach to achieving understanding: the *articulation of systems of interest*; *theory development* for that system; and the deployment of appropriate *methods* to operationalise the theory. The particular and more specific focus of this book is the representation of this knowledge as system *models*.

By adopting a *systems' modelling* focus, as in this chapter so far, it might be argued that an essentially *functionalist* approach is being adopted. That is, the *forms* of organisations and institutions are taken as given and the emphasis is on the way they function both individually and in relation to each other. It is also necessary to explore the deeper structures and forces which create these particular forms of organisation, i.e. to adopt a *structuralist* approach. The position adopted here is as follows. It is argued that a functionalist analysis is usually valuable, at least to provide a framework, and also that this is a useful starting point in comparing contemporary theory with 'classical' urban and regional theory. The approach will be spelled out in more detail in Chapter 3.

KEY IDEA 1.2

A focus on *systems and models* of systems provides some valuable contributions to theory in its own right, and a framework within which other theories can be developed.

We have already noted that there is one discipline which has a prime concern with cities and regions in a holistic way – and that is human geography. To an extent in what follows, 'urban and regional analysis' and 'human geography' can be used interchangeably, albeit with the recognition that this is in respect of the overlapping territories in the Venn diagram which represents the two areas – one multidisciplinary, one a discipline. It is helpful in this context to review briefly the broad stages of the evolution of human geography in this respect. We do this in Chapter 4, in the broader context of all the disciplines which contribute to urban and regional analysis.

The structure of the rest of the book

In the light of the argument thus far, we can now state the objectives of the book through a commentary on the way the remaining chapters are organised. In Chapter 2, we adopt a *substantive systems* focus and try to establish from first principles what the subject matter of urban and regional analysis is. In Chapters 3 and 4, we then complete the framework created to help assemble the pieces of the jigsaw in some order. In Chapter 3, we expand on the three principal dimensions which provide an analytical framework. These were identified in the previous section and enable any piece of work to be categorised. Some concepts associated with system representation and notation are introduced in Appendix 1. This chapter will also, to some extent, help us to provide the basis for analysing the main approaches in the history of the subject. A combination of substantive system definitions in Chapter 2 and the analysis framework of Chapter 3 helps to define the subject matter of urban and regional analysis in a coherent way. In Chapter 4, as a further preliminary, we discuss the conceptual foundations, from a variety of disciplines, needed for effective analysis.

With frameworks established, we can then proceed to the substantive argument and we do this by reviewing in the next three chapters what can be offered as core model-based theory in urban and regional analysis. We adopt, in effect, a *systems* approach. In Chapter 5, we attempt to identify the contributions of classical theorists from the perspective of the framework adopted. This exposes the nature of the analytical tasks and provides a historical context. We relate the implicit 'theory design' decisions of the classical theorists to the frameworks of Chapters 2–4. This is an important exercise for several reasons. It is valuable to understand the origins of some of the foundations of contemporary theory. Secondly, it is interesting to see that, in some cases, if different 'framework decisions' and alternative paths had been explored earlier, the theory – in geography and in the other component disciplines of urban and regional analysis – might have developed more rapidly. Thirdly, much of the classical work is still presented relatively uncritically in contemporary geography texts, and it is important, therefore, to provide a good foundation as a basis for criticism. Chapter 5 provides a relatively brief exposition. A much fuller account is given in Appendix 2.

In Chapter 6, we adopt a modelling perspective. It is only relatively recently that these ideas have been developed to anything like approaching full fruition; and it has become clear that they offer, in many cases, a powerful extension to classical theory and a basis for a general theory of complex spatial systems. In Chapter 7, we show how the classical theory can be rewritten and generalised.

Key ideas are emphasised throughout the book in shaded boxes. It is a particular concern that many of these ideas have not been developed and applied as extensively as would be fruitful. Whilst the research agenda which follows from this observation is at least implicit throughout the book, a more formal summary is presented in Chapter 8.

2

Cities and regions as complex spatial systems

The elements

Cities and regions are made up of an infrastructure carrying the activities of their populations. They are the products of human agents, as individuals, in households or within organisations. These elements combine to form *systems of interest – complex spatial systems*. It can be seen relatively easily that these prime components – people, organisations and infrastructure – can be assembled in a very large number of different ways (at different scales) into systems. Examples are as follows:

- regional systems:
  - cities
  - regions: rural or urban
  - nations
  - groups of nations, like the European Union
- urban systems
  - cities disaggregated to show their structure and workings
- functional systems:
  - economic:
      agriculture
      resource
      manufacturing industry
      consumer services (public and private)
      producer services
  - social
  - labour market
- spatial systems:
  - point patterns (e.g. systems of cities)
  - interactions (commuting flows)
  - networks (traffic, communications)

Cities and regions are systems of a rather special kind: everything at a *place*, traditionally the concern of the geographer. The functional systems involve something more specific such as a 'population' or an 'industry' with the spatial or 'place' dimension identified; spatial systems are more abstract. These systems in different ways form the subject matter of a variety of disciplines. One discipline which above all attempts a synthesis is, as we have seen, human geography, although it inevitably draws on other disciplines. We noted in Chapter 1 the sequence of possible system descriptions that broadly reflects the historical evolution of human geography through four broad phases: from regional to systematic (or functional), to spatial 'science', and to radical and structural. One of the central arguments of this book is that insights from all these perspectives can (and must) be combined in an effective way.

Even this simple framework enables us to build up a portrait of the richness of the subject matter of urban and regional analysis through analytical human geography and associated disciplines. The analysis also obviously connects to a wide range of real-world *problems:* inner city decay, urban transport, rural poverty, regional and national under-development and so on. One of the tests of the effectiveness of the framework to be developed is the extent to which it can lead to fruitful modes of analysis that will contribute to problem-solving in these contexts. The problems can be related to any of the different foci introduced: people, organisations or different kinds of systems. The following are examples:

- people-based:
  - housing quality

- access to services
- access to employment (and therefore income)
- time budgets and lifestyles
- political emancipation
- organisation-based:
  - resource availability
  - technological change
  - market change
  - organisational evolution
- system-based:
  - effectiveness of economic development
  - population, development, resources and famine
  - impacts of economic restructuring (sunbelts and rustbelts)
  - labour markets and unemployment levels
  - urbanisation
  - regional development
  - environmental protection
  - ecological dynamics
  - governmental responses

In practice, the analyst must proceed in specific historical and geographical contexts. Such analysis, however, must be supported by a theory that is as general as possible. The prime objective of the book is to show how modelling can provide at least a framework for such a theory – or, rather, interlocking sets of theories – and the methods which can be used to operationalise such theory in practical situations. We assess the success of these applications in Chapter 8.

The elements in more detail

Here we try to list the prime elements more systematically, extending the people–organisations–infrastructure notions introduced in the previous section. We examine the ways in which the elements can be grouped into systems of interest. We also focus on the relationships between the basic entities. As a preliminary, however, it is necessary to consider the concept of 'level of resolution' or 'scale' – extending the discussion that was initiated in Chapter 1.

It is clear from the notion of 'level of resolution' associated with microscopes or of scale

associated with maps, that the finer the resolution level, the more detail is on offer. In principle, the maximum possible amount of detail is desirable; but there are many circumstances where it cannot be achieved because of inadequate data; and more importantly, as we shall see, there are different phenomena which can be recognised and studied theoretically at different scales. Indeed, some of the hardest theoretical questions arise from relating descriptions and theories at different scales – the so-called *aggregation problem* (which is well recognised as such in economics) to which we shall return frequently in one guise or another.

The urban and regional analyst faces level-of-resolution questions in at least three ways: entities which are components of systems of interest have to be defined and *categorised*; many of them have to be *located* in space; and their behaviour has to be described over *time*. The decisions on these issues therefore have three aspects: *sectoral* (number and breadth of categories), *spatial* (size of area units within which entities are to be located) and *temporal* (length of time units which provide the basis for longitudinal description and analysis). For example, a population geographer might characterise the population of a region by the three categories of age, sex and employment (defining 'units' within each of these categories, e.g. 0–4 years, 5–9 years and so on) but neglecting all other possible categorisations. He or she might relate them to locations based on local authority units (perhaps so that Census data can be used). Ten-year time intervals might be used (again because of restrictions caused by the availability of data).

Partly because of the vast range of possible choices of scale for particular pieces of analysis or theorising, and in part because we may be selective in relation to which elements to focus on in a system of interest, it is clear that 'systems of interest' may be defined in many different ways. We should not be seeking uniqueness of definition, but simply definitions that are effective for particular purposes. We proceed in a pragmatic way, indicating where detail is likely to be important, but also recognising the value and necessity of more coarse scales on occasion.

We begin by breaking down the three kinds of elements introduced in the last section. We distinguish the *products* of organisations from the organisation itself; and we distinguish *land* as a special component of infrastructure. The basic entities with which the urban and regional analyst deals can be identified broadly as follows:

- people
- organisations
- commodities, goods and services
- land
- physical structures and facilities

For 'people' and 'organisations' in particular, we can distinguish

- activities and processes

and for all entities:

- relational structures

We discuss each of these briefly in turn.

People can be characterised by age, sex, education, position in a household, job description and location, income, wealth, house type and residential location, shopping frequencies, baskets of goods purchased and locations, recreational activities and locations, services used and locations, and allocation of time amongst different activities. Many other characteristics could be added, of course, but this list will serve as an illustration for the time being and emphasis has been given to geographical aspects by noting the *locations* for a variety of activities. The types of 'problems' of people broadly sketched earlier can then be discussed in a more interesting way by relating them to these categories. Housing problems, for example, are much more likely to be acute for unemployed, elderly or low-income people. The specification of categories is itself a task for the theorist: there is no absolutely 'right' way to do it. But any account of the population in categories provides a statement of 'initial conditions' and two key questions for the theorist: how did that state come about? How will the system evolve?

Organisations are described by their role and purpose, a list of people involved in them in different functions, their activities (such as production of goods or services) and their cash flows. Classification systems are available, such as the SIC (*Standard Industrial Classification*) in the UK which, in principle, covers all kinds of

organisations – though in practice, there are biases, for historical reasons, towards manufacturing. The notion of an organisation is intended to be very broad at this stage.

It is worthwhile even at this early stage in the argument to attempt to describe the rich variety of organisations which constitute economic systems. It is worth trying to add a lot more detail even at this stage. An indication of this variety is therefore given in Tables 2.1–2.9. The information in these tables builds from coarse to finer levels of sectoral resolution. Unfortunately, there is no standard way to present this information. Census tables are usually geared to the *Standard Industrial Classification* (or its post-1981 equivalent in the UK, the industrial group). This was designed when the economy was dominated by manufacturing (and we have therefore used it in Table 2.3) much

---

Table 2.1 Types of organisation

Primary
Manufacturing
Industrial and domestic services
    Utilities
    Construction
    Transport
    Communications
Professional and scientific
    Financial
    Legal
General governmental services to the community
    Administration
    Defence
    Justice
    Other
Housing
Distributive trades
    Wholesaling
    Retailing
Personal services
    Education
    Health
    Social
    Fire
    Police
    Miscellaneous
Cultural and recreational

more than at the present time. Accordingly, these headings have been ordered differently, and sometimes subdivided, in the following tables. We need to recognise at the outset, of course, that many large organisations are multi-sector conglomerates which need, correspondingly, multiple classification entries.

The overall rationale for the scheme is presented in Table 2.1 at the coarsest level of resolution. The first two headings are the standard ones of 'primary' (mainly agriculture) and 'manufacturing', with the latter usually taken as the 'secondary' sector in the economy. These are further subdivided in the more or less standard way in Tables 2.2 and 2.3. The remaining headings of Table 2.1 subdivide the service sector in a way which is not standard but which seems to bear more resemblance to its importance in present-day geographical and economic structure. The first such heading is 'industrial and domestic services' and this covers the supply of key inputs both to organisations and to residences. The sub-headings are self-explanatory and are listed in more detail in Table 2.4. They range from highly industrialised services through to various kinds of professional services but they are grouped together here because of their common function in supplying services to both individuals and organisations all of which play an 'intermediate' role of some kind. Because in many of these cases, the main customers are other organisations, these services are often now known as *producer* services. This is followed by another class of general services, this time supplied by government, but of a type which does not usually impinge directly on the individual. These are listed in more detail in Table 2.5.

All the remaining headings are presented together at the end of the list because they mainly represent the organisations which, in various ways, meet the needs and support the activities of individuals and households. These are mostly, therefore, 'personal' or 'finally consumed' services.

Table 2.2 Primary sector

Agricultural
Market gardening
Mining
   Coal
   Ores, etc.
Oil

Table 2.3 Manufacturing sector

III	Food, drink and tobacco
IV	Chemicals and allied
V	Metal manufacture
VI	Mechanical and electrical engineering
VII	Shipbuilding, etc.
VIII	Vehicles
IX	Metal goods
X	Textiles
XI	Leather
XII	Clothing
XIII	Bricks and pottery
XIV	Timber and furniture
XV	Paper, printing and publication
XVI	Other

Table 2.4 Industrial and domestic services

Construction (excluding housing)
Energy
   Gas
   Electricity
   Coal
   Oil, etc.
Water supply
Sewage
Waste disposal
Transport
   Road
   Rail
   Air
   Water
Communications
   Postal
   Telephone
   Cable, etc.
   Broadcasting
Professional and scientific
Financial
   Banking
   Insurance
   Finance

Table 2.5 General governmental services

Administration
   Departmental
   Inland Revenue, etc.
Defence
   Army
   Navy
   Air Force
Justice
   Courts
   Prisons

Table 2.6 Housing

Construction (housing)
   Private
   Local authority
   Housing association
Building societies
Local authority administration

Table 2.7 Distributive trades

Wholesaling
Stockholding
Import/export
Retailing

Butcher	Garage
Baker	Car sales
Greengrocer	Chemist
Fishmonger	Car accessories
Grocery	Photography
Dairy	Take-away food
Specialist foods	Wine
Supermarket	Optician
Hairdresser	Florist
Tailor	Toys
Clothing	Antiques
DIY	Auction room
Cosmetics	Haberdasher
Dry cleaning	Music/records/video
Off-licence	Jewellers
Tobacconist	Leather goods
Launderette	Fancy goods
Confectioner	Office equipment
Footwear	Department store
Ironmonger	Bookshop
Radio/TV	Art supplies
Electrical	Sports goods
Hardware	Betting shop
Furniture	Newsagent
Soft furnishings	

The first such heading is concerned with housing supply, and some of the organisations involved are listed separately in Table 2.6. A wide variety of personal goods and services are supplied, mainly through the private sector, by what we have broadly called retailing organisations, a long list of which, to illustrate the variety, is presented in Table 2.7. The remaining headings relate to the use of mainly public services and facilities of various kinds, though the definition of what is normally construed as 'public' and what as 'private' has changed dramatically in the UK and elsewhere in the 1980s and 1990s. We distinguish first those supplied by various organs of government (Table 2.8) and then, under the heading of 'cultural and recreational amenities' (Table 2.9), another wide range, some of which are supplied by the public sector, some by the private sector.

It is clear from Table 2.8, however, notwithstanding the qualification about public–private demarcation, that the various headings cover much of the main expenditure of government agencies at both national and local scales and so form a major part of the 'public sector'.

An alternative perspective on organisations is to look at stock market classifications and Table 2.10a shows the recent London classification as listed in the *Financial Times*. This gives a different kind of portrait of the structure of the private sector. It is interesting that at the time of completing the writing of this book, the *Financial Times* has adopted yet another classification which is shown as Table 2.10b.

Commodities, goods and services are the products of organisations. They correspond roughly to the usual categorisations of an economy as concerned with primary, secondary and tertiary production, though this is only a rough correspondence which needs elaborating later. Sales can be to consumers (usually via wholesalers

Table 2.8 Personal services
Education
Nursery schools
Primary schools
Middle/secondary schools/sixth form colleges
Colleges of further education
Technical colleges
Colleges of higher education
Universities
Health
Hospitals
General practitioners
Community services
Dental services
Public health
Ambulance services
Social services
Social security
Child care
Other welfare
Probation, etc.
Fire
Police
Miscellaneous
Refuse disposal
Cemeteries, etc.

Table 2.9 Cultural and recreational
Indoor
Museum
Art gallery
Exhibition hall
Theatre
Cinema
Concert hall
Library
Public houses/bars
Clubs
Churches
Amateur societies
Political parties
Outdoor
Allotments
Sporting facilities
Cricket
Soccer
Rugby
Bowls
Skating
Hockey
Swimming, etc.
Recreation
Walking
Riding
Sailing
Flying
Motoring
Golf
Fishing
Bird watching, etc.
Children's play areas

and retailers) or 'business to business', through what can be quite complicated supply chains. The notion of interaction is therefore again critical – in this case inter-sectoral, which is captured in input–output models (which we will pursue in the context of regional economies later).

Land is distinguished (as a commodity, or what economists might call a factor input) because of its special significance to geographers: an area of land is a *place* and in that sense a basic unit of study. It is also unlike most other commodities and products in an important respect: its supply is largely fixed and given and cannot be expanded or substituted.

Physical structures and facilities refer to all buildings, for example, and to other kinds of infrastructure like roads, rail systems and airports. All forms of communications are obviously of increasing importance in this context.

Activities and processes can be applied as concepts to both people and organisations. For people, they have in the main been listed among the descriptions of lives in the appropriate paragraph above. For organisations, these terms represent the means whereby goods and services are produced. Activities, and the interactions between them, usually mediated across space, are critical to theory-building for model development.

It is already clear from the definitions which have been introduced so far that no entity functions alone. The idea of *relational structures* is that it is possible, in principle, to list the relations of any one entity with any other. The interaction relating consumer activities and retailer activities provides

Table 2.10a The old London stock market classification of organisations

Alcoholic beverages
Banks, retail
Breweries, pubs and restaurants
Building and construction
Building materials and merchants
Chemicals
Distributors
Diversified industrials
Electricity
Electronic and electrical equipment
Engineering
Engineering, vehicles
Extractive industries
Food producers
Gas distribution
Health care
Household goods
Insurance
Leisure and hotels
Life assurance
Media
Oil, exploration and production
Oil, integrated
Other financial
Paper, packaging and printing
Pharmaceuticals
Property
Retailers, food
Retailers, general
Support services
Telecommunications
Textiles and apparel
Tobacco
Transport
Water

Table 2.10b The April 1999 *Financial Times* London stock market classification of organisations

Aerospace and defence
Automobiles
Banks
Beverages
Chemicals
Construction and building materials
Distributors
Diversified industrials
Electricity
Electronic and electrical equipment
Engineering and machinery
Food and drug retailers
Food producers and processors
Forestry and paper
Gas distribution
General retailers
Health
Household goods and textiles
Information technology hardware
Insurance
Investment companies
Investment trusts split capital
Leisure, entertainment and hotels
Life assurance
Media and photography
Mining
Oil and gas
Packaging
Personal care and household products
Pharmaceuticals
Real estate
Restaurants, pubs and breweries
Software and computer services
Speciality and other finance
Steel and other metals
Support services
Telecommunications services
Tobacco
Transport
Water

an obvious example. At this general level, it serves to remind us that such a very complicated list underpins any system of interest – and that understanding them is critical to understanding the interdependencies in the system. It also serves to help us to *define* systems in the more formal sense in which the term is used in 'systems theory' or 'systems analysis': a system in this more technical context is defined as a set of elements which are interdependent in some coherent way. A retailing system might be defined as made up of shops, suppliers of goods to shops (manufacturers and wholesalers) and consumers (distributed in their residences and workplaces which form the origins of their shopping activities). Subsystems defined in this way are not, of course, wholly self-contained, but they provide a useful basis for study. We should also note the relationship of this argument to scale. At the finest scale, an 'agent' is a 'subsystem' – and this argument then reminds us that we should consider the behaviour of the agent in relation to the structure provided by the wider environment. The agent's behaviour is partly determined by this structure; the structure can, to an extent, be modified by the agent. This kind of relationship, the agency–structure problem, will recur in various guises as the argument unfolds. In the next section, we add some flesh to the bones of this rather abstract presentation by considering examples.

Urban and regional systems at different scales: examples

To fix ideas, we need to think concretely about the wide range of possible systems without being swamped by detail. We can take the argument forward in two stages: first in relation to whole systems at different scales; and then a variety of functional systems which combine elements according to some primary focus. The latter set are essentially overlapping subsystem models. Our examples will include agricultural and industrial systems. These are treated as distinct, but they overlap, for example, through mutual suppliers (e.g. commercial vehicle manufacturers).

We start by looking at whole systems. The boundaries of urban and regional systems can be shown on maps at different scales. We can also assume, as is common practice, that the main infrastructural features of each of these systems (like road and rail networks and urban settlements) can be shown on these maps. Examples include the following:

- a system of countries, e.g. the European Union – a broad-scale inter-regional system
- a system of regions within a country, e.g. the regions within the UK
- a system of cities within a region
- a city region and the structure within it

The main differences are of scale. What is usually represented on familiar maps is the infrastructure. It is not hard, however, to imagine the complex set of activities carried by this infrastructure.

There is, potentially, a long agenda of (sub)systems of interest for study. The aim is to work with a sufficiently varied set of systems in this book to establish the principles that will enable the reader to be able to connect to the wider literature and to construct a similar argument for himself or herself in relation to any urban and regional system of interest.

The examples we will pursue in detail are as follows:

- agricultural systems
- industrial systems
- residential location and housing
- service delivery systems – of which retailing will be the archetypal example
- transport systems
- we can then return to integrated systems with the twin aims of (i) handling interdependence of subsystems and (ii) being able to represent the main elements of urban structure. The subsystem models can be components and we have a much richer starting point.

At a broader scale, we will also be concerned with

- demography
- economic systems

In order to be able to approach the system modelling task effectively, we next seek to develop an appropriate framework for analysis.

3

A framework for analysis

Three principal dimensions

In Chapter 2, we attempted to enumerate the basic elements which can be combined in various ways into urban and regional systems and subsystems and to emphasise our prime concerns with spatial relationships – interaction and location. Chapman (1977) remains unusual among geographical writers in emphasising the importance of enumerating elements, calling it *entitation*. Here, we extend these notions and identify three principal dimensions, each of which subdivides and represents a set of decisions to be taken by the analyst. These then provide the framework against which any piece of theory or analysis can be developed (for research design purposes) or understood and assessed (in reading the literature).

The three dimensions can be called

- system articulation
- theory
- method

We have already argued that there are many ways of defining *systems* and that the key task is to choose an effective definition for a particular purpose. We can add to the notions of entitation and levels of resolution and we do this in the next section. The key enterprise for any urban and regional analyst in relation to any system of interest is the *understanding* of that system; and it is this understanding which is represented as *theory*, the second of our principal dimensions. The third dimension is *method*, i.e. the set of tools for operationalising a theory. The relationship between theory and method can be very close and

in some cases (as we will see in the case of so-called 'classical' theory below) good theory was inadequately developed because appropriate methods were not available or not chosen. This interplay of theory and method helps us to interpret the history of urban and regional analysis and closely related disciplines like geography and to pinpoint those research problems where rapid development could be expected if certain methodological problems were solved.

For short, we will call the three principal dimensions the STM dimensions (for system–theory–method). We discuss them in turn in the following sections.

System articulation

INTRODUCTION

The three main sub-dimensions in defining a system of interest are as follows:

- entitation
- levels of resolution, with three sub-components:
  - sectoral
  - spatial
  - temporal
- spatial representation

We have already discussed and illustrated the first two in working towards system definitions in Chapter 2. Here we focus on the third, which is implied by the 'spatial' aspect of level of resolution but which is so basic to urban and regional analysis – and human geography – that it deserves extended treatment.

SPATIAL REPRESENTATIONS: CARTESIAN CONTINUOUS SPACE VERSUS DISCRETE ZONES

In order to develop a *geographical* theory, we need an underlying spatial framework on which to locate the entities. There are various ways of achieving this – in particular using continuous space (say using Cartesian co-ordinates as in the Ordnance Survey Grid) or discrete space (a zoning system at some scale). The differences in representation turn out to be very important. Much traditional geographical theory is based, disadvantageously as it turns out, on the continuous-space representation.

It is difficult to imagine a form of urban and regional analysis which does not involve at least an implicit spatial representation. If, simply, two 'places' are being compared for instance, then the boundaries of those places will have to be defined in some way. If 'inner city' problems are to be discussed, then the extent of the inner city has to be defined. And so on. More typically, as argued in Chapter 1, we will be concerned with the *location* of entities or the flows (the spatial interaction) between locations. We therefore need to *represent* location, i.e. to have a spatial representation.

There are two common ways of proceeding, each method having possible variants; the two can also be used in combination. The first is to measure the location of an entity, say a firm, by its Cartesian co-ordinates as shown in Figure 3.1a; the second is to impose a discrete zone system and to measure location by specifying the zone in which the entity is located, as in Figure 3.1b. The second obviously represents a more coarse level of resolution than the first (or the first can be considered as a limiting case of the second, as the number of zones is made larger and their size smaller). In general, in the second case, the smaller the average size of the zone, the finer is the spatial scale.

Discrete zone systems can obviously take many different forms, both in relation to level of resolution as we have seen; and also in relation to the shape of the zones. The tidiest, and often most convenient, system to use would be the kind of square grid shown in Figure 3.1c. However, quite often we are forced to use administrative units, such as the wards of a city (as shown in Figure 3.1d), because the only available data are for such a system.

Two obvious examples which illustrate the continuous and discrete representations from classical theory are the simple Weber problem and the elementary gravity model, shown in Figure 3.2a and b respectively. The second of these examples also illustrates a feature common to many discrete zone systems: the flows (in this case) are considered to be between points – the centroids of each zone – so that geographical analysis using a discrete zone system is often equivalent to locational analysis on a set of fixed points; these, in turn, are often referred to (in the operational research literature, for example) as fixed points *on a network*. This nomenclature is also potentially confusing because it may also be appropriate to relate a discrete zone system to a real (say road) network, and the zone centroids have to be connected to the network by dummy links as shown in Figure 3.3.

It is convenient to summarise the argument so far about spatial representation and then to add one element to it. Our primary concerns have been with location and interaction. The first can be handled (as seen in the figures) in either representation. The second has been exhibited for a discrete zone system and we have shown in principle how an underlying network can be added. Flows can be represented also in a continuous space representation – but flow *densities* would have to be used and network-type information would have to be handled through something like velocity fields. This approach was developed by Angel and Hyman (1970, 1972, 1976) but has not been extensively used since.

As the whole argument develops below, we will see that with currently available methods, the discrete zone system will, on the whole, be more convenient – both in theory and practice. But it is essential to be aware of the distinction and, as one basis for critical appraisal, to be able to investigate the assumption about spatial representation, implicit or explicit, which has been made in any particular piece of geographical analysis.

The discrete zone system, and the algebraic notation used to describe it, turn out to be very important in the development of urban and regional theory. For readers who are unfamiliar

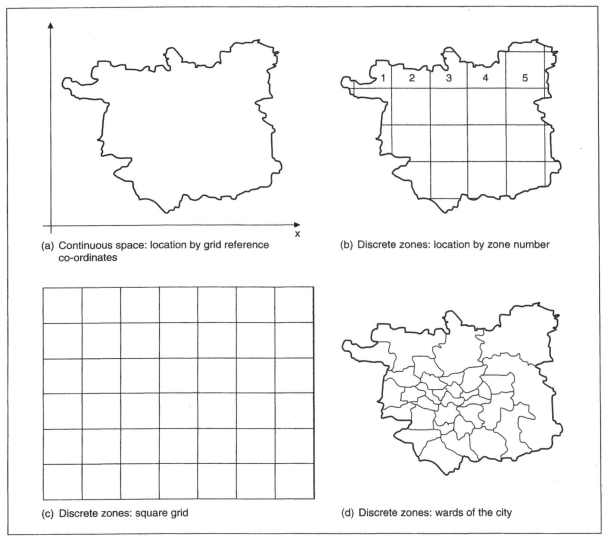

(a) Continuous space: location by grid reference co-ordinates

(b) Discrete zones: location by zone number

(c) Discrete zones: square grid

(d) Discrete zones: wards of the city

Figure 3.1 Alternative spatial systems

with it, an introduction is provided in Appendix 1, in the context of developing some interesting ideas which arise from having a good notation.

BOUNDARIES AND CATCHMENTS

To complete the argument, we have to add the geographer's concern with *boundaries*. It may be appropriate, for example, to delineate the catchment area of a school or the market area of a shopping centre. Or we might have a theory about the location of administrative boundaries. At first

sight, it appears that the continuous space representation handles these questions better because the boundaries can be drawn precisely. In the discrete zone case, whole zones would have to be designated 'inside' or 'outside', and the resulting boundary would then be a composite of zonal boundaries which would be only an approximation to the precise boundary. This is less of a problem if small zones are used. These situations are exhibited in Figure 3.4.

There is one complication for this part of the argument. Many boundaries (e.g. for market areas

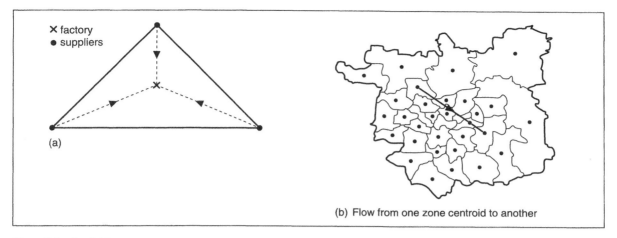

(b) Flow from one zone centroid to another

Figure 3.2 Spatial flows

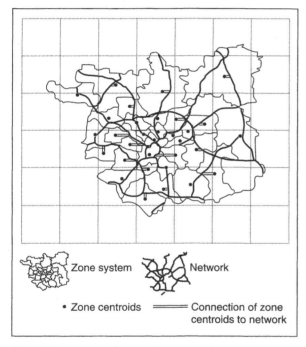

Zone system   Network

• Zone centroids   Connection of zone centroids to network

Figure 3.3 A zoning system with nodes and network

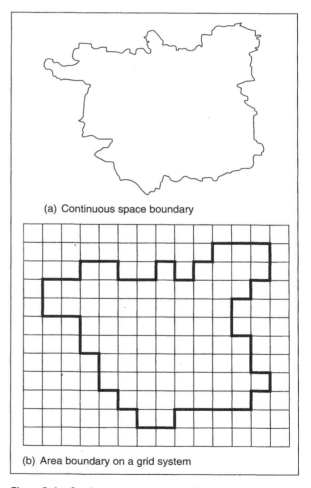

(a) Continuous space boundary

(b) Area boundary on a grid system

Figure 3.4 Continuous space versus grid systems

or catchments) are not precisely defined because of overlapping flows. In the continuous space case, this can be handled using probability contours (Figure 3.5a), in the discrete case by having a mix of patronage in a zone (Figure 3.5b). It turns out that these concepts in the discrete zone case can be used to give a much more precise definition of catchment populations than is commonly used even now. These ideas are used to illustrate the notation in Appendix 1.

Further, we note that where administrative boundaries are involved in analysis (e.g. because of data) it is usually possible to devise the zonal boundaries so that they are contiguous with the administrative ones where appropriate. (There is a problem, however, when several data sets are used, each relating to different and non-matching boundaries.)

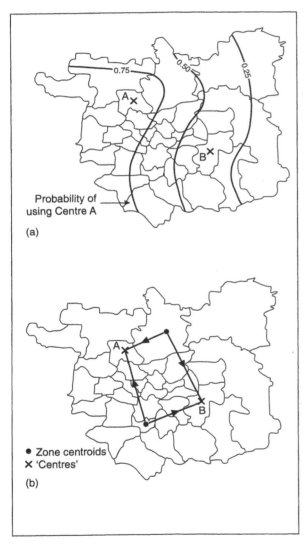

Figure 3.5 Zone centroids and 'centres'

CONCENTRATED AND DISPERSED ACTIVITIES

Before concluding this section, it is useful to add a note about a distinction which can usefully be made between different types of activity with respect to space consumption because this has a bearing on the spatial representation that is used in the different cases. We follow Paelink and Nijkamp (1975) in distinguishing *concentrated* and *dispersed* activities. They also distinguish consumers and producers, which virtually parallels our own distinction between people and organisations. The only difference between the schemes is that consumers may sometimes be 'other organisations'. They draw up the four-way classification shown in Table 3.1. Consumers can be considered to be either dispersed or concentrated; likewise producers. The relationships between them can then in principle take one of the four forms shown.

Table 3.1. A spatial classification of activities

	Producers	
	Concentrated	Dispersed
Consumers		
Concentrated	i	iii
Dispersed	ii	iv

Dispersed activities can be considered as 'space consuming' and concentrated activities as taking place at a point in space. The distinction is not, of course, as sharp as this. All activities consume space and there are cases where the competition for land among so-called concentrated activities is crucial, as we will see later.

We can find examples in relation to the population and the organisations of Table 3.1. The housing of the population is an obvious example of a space-consuming activity, and the other obvious one is part of the primary sector – agriculture. Most of the other activities, as a first approximation, can be considered to be concentrated. Let us now consider examples of the four cases.

(i) Both consumers and producers are essentially points. This covers relationships between organisations, e.g. the spatial connections of a firm to its suppliers and to the wholesalers who handle its products. But these matters are also determined by spatial scale. At a coarse scale, we might be interested in migration flows, or flows of goods between towns; then the towns can be considered as points.

(ii) In the second case, consumers are dispersed and producers are concentrated. This covers a wide range of examples where the consumers are people distributed among their residences and viewed at a reasonably fine spatial scale. The 'producers' would then be places of employment or providers of any of the wide range of services used by the population. Paelink and Nijkamp call this the 'market area' case, since the population can be grouped into (possibly overlapping) market areas around producers for different purposes.

(iii) The third case, concentrated consumers and dispersed producers, is well illustrated by farmers (whose activities are dispersed) in relation to their markets, say towns, which can be considered as points.

(iv) The dispersed–dispersed case is at first sight more difficult to illustrate. Paelink and Nijkamp cite the example of 'megalopolis', a dispersed urban region, as in Gottman's (1964) treatment of the North-East Corridor of the United States. However, if we consider the population in relation to services which seem correspondingly dispersed, like small shops, then we might consider this a better example of this category. However, this fourth case shows that the classification is only approximate.

We can handle either kind of activity in either discrete or continuous space. For concentrated activities, the 'point' will either 'move' continuously as in the Weber problem, or be located at one of a fixed set of points (which can be considered as zone centroids). For dispersed activities, the analysis problem in continuous space is the identification of boundaries which demarcate different land uses – as in von Thunen's problem – or the categorisation of zones by land use type in the discrete case.

Theory

INTRODUCTION

We begin with a brief discussion of the nature of theory at a very general level and then we examine, still in a broad way at first, the nature of *urban and regional* theory. It is difficult, and probably not particularly helpful, to define precisely what is meant by a 'theory'. The notion has different colloquial meanings and this does not make this task any easier. For the purpose of this book, a rough working definition will suffice and readers interested in taking this particular argument any further can delve into the literature of the philosophies of science and knowledge.

TRUTH AND UNDERSTANDING

There are at least two concepts which should form part of a working definition: one is 'truth'; the other 'depth of understanding'. In developing a theory about some system of interest, we are seeking to explain why it is as it is and how it functions, how it has evolved, and how it might develop. We are clearly seeking an *explanation*, but also one that is true. A theory is tested by comparing its predictions with observations. A relatively untested theory will be called a hypothesis; a heavily tested and well-established one, a law. We use 'theory' as a general term to cover the whole spectrum of cases.

The second notion of *depth* of understanding or explanation has to be coupled with that of 'truth'. It may be relatively easy in a particular study to be trivially true, e.g. with a piece of elementary description. This may be important in itself, but if it is deemed to constitute a theory, that theory will be at the lowest level. So the more ambitious a theory is in terms of depth, the more it 'says'; and it may be less likely to be 'true'.

What indicates that a theoretical explanation is being attempted at a higher level (or greater depth) is the introduction of concepts which are more abstract than those involved in the definition and description of entities which are the direct subjects of observation. This will be elaborated through examples in the rest of the book.

The notion of 'true theoretical explanation' is made more difficult when we add questions of 'how true?' to questions of depth. Surely, it might be argued, something is either true or it isn't? Well, the idea of a theory being 'partly true' can arise as follows: the explanation offered may be recognisably true in some broad respects, but not in detail. It may be argued, for example, that residential densities decline with distance from the central business district of a city and theoretical reasons may be offered for this. Observations may show that, if density is measured in terms of broad averages relating to equally spaced concentric rings around a centre, this is indeed the case. But a more detailed examination may reveal pockets of high-density housing in outer parts of the city. How do they come about? This raises new theoretical questions without necessarily taking away the *insight* offered by the overall theory. We might then know that we have some of the building blocks of a good theory, but that the final theory will be more elaborate. It may turn out that this is a modified form of the original one, or alternatively that an entirely new formulation is needed using new (probably in some sense higher level) concepts.

Thus, we are seeking the greatest possible depth of understanding, and we are seeking truth; but it is often necessary to be tolerant of the shortcomings (less than the whole 'truth') of a theory because it does offer important insights in spite of them, and because it is the best that has been managed to date. Such theories provide platforms for further research. *An implicit theme in the rest of the book is that a systems modelling approach provides such platforms.*

THEORY IN URBAN AND REGIONAL ANALYSIS

We can now begin to discuss theory in urban and regional analysis, and we do so first in terms of the two most general kinds of geographical questions: location and spatial interaction respectively, beginning with location. This can be approached in two ways: first, it is possible to focus on any activities (or associated physical infrastructure) and to examine their spatial distributions; or secondly, it may be appropriate to focus on an area and study which activities are located there. This can be seen very clearly using the device that was introduced in a review article by Lowry (1967).

Assume a geographical region is divided into a number of zones in the usual way. Let the zones be numbered 1, 2 … (Recall or reread Appendix 1 if appropriate!) Let the different kinds of activities also be labelled 1 (say, residential), 2 (manufacturing), 3 (retailing), … and so on. Now consider Figure 3.6. In each square of the figure, we can enter, in suitable units, the amount of that activity which takes place in that zone. If we read along rows – say one particular row – then we are focusing on distribution of that activity across the region. If we read down a column, we are focusing on the mix of activities in a particular zone. Since all the activities consume land, an appropriate common set of units would be amount of land used, and then the whole table gives a detailed account of land use by the different activities across the region. A focus on rows emphasises the *location* of particular activities; a focus on columns emphasises *land-use mix* at a location.

The fact that the same people are involved in different activities in different places, and that organisations in different places have connections (e.g. in commodities being delivered as an output from one and an input to another) means that the system can only function if there are *flows* between

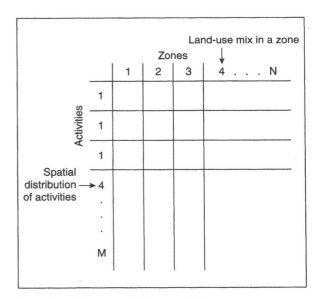

Figure 3.6 The Lowry tabular presentation

locations: people going from home to work, shop, hospital, school; goods flowing between organisations; and so on. A second major element of geographical theory, therefore, is *spatial interaction*. If the activities of the organisations that are listed in Tables 2.2–2.10, and those which are familiar to people, are enumerated then, typically, there can be a flow from any one to any other. Some examples are provided in Figure 3.7. In effect, we are identifying the location of a number of sample activities and noting that people or goods have to move around or be moved around as part of the functioning of the regional system as a whole.

In the figure, there are movements from residences to all other 'activity-locations' for various reasons: to other residences for social and domestic purposes; to all other activities for people who work there; to service locations for people to use those services (or occasionally, the service might come to them, in the form of an ambulance, for example). There are flows of goods from manufacturing to all other sectors (though the flows to consumers will usually be first via wholesalers and then retailers). There are links between workplaces and educational establishments for training purposes; and between government administrators and all activities in relation to government grants, health and safety legislation, and so on. These examples serve to illustrate the very rich variety of spatial interaction. In many cases, such flows represent manifestations of the relational structures introduced in Chapter 2.

The basic problems of urban and regional theory can now be seen to be concerned with the location of activities of people and organisations and associated physical infrastructure and with the spatial interactions between the activities at the different locations. The way in which these questions have been developed above already begins to give a glimpse of how they can be interesting and difficult, especially when different subsystems are interdependent. Two or three examples will suffice: first, different activities compete for land, and because spatial interaction is of differing degrees of importance in different cases, land will come to have different 'values' because of this competition and because of the

Residence	to	work
Residence	to	shop
Residence	to	school
Residence	to	hospital
Supplier	to	manufacturer
Manufacturer	to	wholesaler
Wholesaler	to	retailer

Figure 3.7 Interactions

relative accessibility of locations; secondly, spatial interaction can only be carried on the physical networks provided, and these also are competing for land; thirdly, we can begin to see that a theory of flows can be built on a knowledge of the location of activities, but also vice versa!

THE PARTIAL–COMPREHENSIVE SPECTRUM

There is one other preliminary to be tackled: whether we seek to achieve a partial or a comprehensive approach to theory-building. In effect, we have defined problems for analysis – for theory-building – in relation to location and spatial interaction patterns for particular subsystems. And we have also begun to see that there will be more comprehensive problems when we recognise the interdependence of subsystems (as through the competition for land).

Consider now an industrial sector within a city. One geographical problem is to investigate the locational behaviour of a single firm, *given* the rest of its own sector and the structure of the city. This is a highly *partial* approach. If we attempt to understand the joint behaviour of all the firms in the sector, this is a much more *comprehensive* (and much more difficult) problem. If we try to understand this joint behaviour simultaneously with other features of the city, the approach is yet more comprehensive. There is clearly a partial–comprehensive spectrum and it is useful to know where on such a spectrum any particular piece of analysis stands. It is a valuable distinguishing feature.

There is a fundamental difference in partial and comprehensive approaches – and there are different degrees of 'partialness'. The simplest kind of problem involves the locational or interaction behaviour of one element – a household, a firm, a service unit or whatever – *given* the rest of the system which provides the environment for the decision involved. The approach becomes less partial as more units are made endogenous and can interact with each other. This may happen within a sector: we may try to model all firms of a particular type which are competing with each other – still *given* everything else. Or we may consider a population of households competing for residential accommodation, but again given the distribution of jobs and services and the other geographical variables on which the residential location decisions might be based. The element of competition draws in new features to the theory: particular groups of people may be attracted to each other; other groups repelled. These features are independent of the other variables, but will affect the final spatial distribution and the nature of the rent surfaces to be predicted in a theory. In the most comprehensive approach, units in different sectors interact with each other.

The more comprehensive an approach becomes, the more interdependence has to be dealt with explicitly by the theorist. In the end, a successful theory has to be comprehensive; but partial approaches are valuable in two ways. First, they are often useful in themselves as good approximations. Secondly, they are the building blocks of more comprehensive theories.

THE UNDERPINNING OF THEORY: THE BEHAVIOUR OF 'AGENTS'

The next stage of the argument is to attempt to build explanation by examining the behaviour of the main agents involved in particular subsystems with the aim of achieving an understanding of this behaviour which can then provide a foundation for theories about location and interaction patterns and the ways in which these change. In other words, we have to try to identify the *component hypotheses* from which we can start to build up a more comprehensive theory. (We will also be helped in this exercise in the next chapter when we look at what a range of disciplines can offer.)

We have, implicitly, identified three main types of 'agents', making choices about their activities (though 'choices' which are constrained, often heavily, by the environment in which they are made). These agents are listed in the tables of Chapter 2. There are people, whom we now sometimes refer to as *consumers* of goods and services; there are entrepreneurs and employees; and there are politicians and civil servants in the public sector – the agents of government. In many cases, the distinctions become rather blurred, but they will suffice for the time being.

At the micro level, geographical theory must be rooted in theories of the behaviour of these different kinds of agents. As we will see in more detail in Chapter 4, when we discuss the contribution of economic theory, we look mainly to the theory of consumers' behaviour on the one hand and the theory of the firm on the other. It is frequently assumed that consumers behave in such a way as to most nearly achieve their preferences subject to constraints such as availability of income; and that entrepreneurs act to maximise the profits of the firms they own. There is less consideration directly in the economic literature on the role of government agents. In all cases, it can be immediately recognised that life is more complicated than simple theory suggests. The essential task for the geographical theorist, therefore, is to select the features of available theory which seem to be most important for geographical problems – though this is easier said than done.

The overall framework can be described simply. People have, broadly, four sources of income: capital, the ownership of wealth or assets, with associated income from interest or rent; wages or salaries associated with work; 'savings', e.g. in the form of pensions; or payments from government agencies, such as unemployment and child benefit. Organisations provide goods and services both to people ('final demand') and other organisations ('intermediate demand'), either at market value with the firms aiming to maximise profits, or to meet a need at prices determined by the costs of factor inputs and whatever can be 'agreed' through the political process about the level of taxation. All these organisations provide jobs which are the sources of the wages and salaries. From this simple basis, we explore a sample range of problems for urban and regional theory in the next section.

INTERDEPENDENCE

One of the crucial aspects of urban and regional theory is the interdependence of phenomena at different scales. We can usefully distinguish at least the macro, the meso and the micro. At the *macro* scale, we might be concerned with the population and economy for the whole of a study region, but with little or no spatial differentiation. (There might, however, be a good deal of sectoral or temporal resolution.) At the *meso* scale, space is differentiated at least by coarse zonal units such as those shown in Figure 3.1, though these units may be quite small. At this scale, however, we usually add a sectoral condition to the spatial one: we assume that it is too coarse for individuals or firms to be identified. At the *micro* scale, these individuals or firms can be identified, and we assume for our purposes that there is a sufficiently fine spatial level of resolution in use as well.

We noted above that individuals and entrepreneurs, for example, are assumed to make decisions about their activities and locations, usually subject to constraints of various kinds, in what are seen to be their own best interests. What kinds of variables could play a role in such decisions? An individual or household making a residential location decision will take into account the quality of the environment of alternative locations together with the access they provide to various kinds of facilities (jobs, schools, services and so on). The entrepreneur will take into account his or her potential market, labour supply, accessibility to factor inputs, the cost and availability of land, and so on. There will be many other variables in each case, of course, but the examples above each have one feature in common: they are determined in aggregate by the corresponding decisions of other individuals and entrepreneurs, and not simply those acting in direct competition.

Residential density, for example, is determined by the collective past decisions of individual residents and of suppliers of housing. Access to facilities is determined by the previous decisions of entrepreneurs and suppliers of public services. The market area (for sales and for labour) for entrepreneurs is determined by residential location decisions, and by the location decisions of other entrepreneurs. All these variables are properties of locations which are either averages for the location (e.g. density) or functions of the whole surrounding environment (e.g. accessibilities). At any one time, locational decisions will be taken given the current values as a backcloth to the decisions. However, the backcloth will then change marginally, and occasionally rapidly and substantially, as the decisions are made. These kind of links and interdependencies will be the stuff of urban and regional theory.

SUMMARY

We can summarise this part of the argument by recalling that the urban and regional analyst is concerned with the *pattern* of activities and associated infrastructure and flows (at different scales), and also with the *processes* that are continually changing this pattern. (For clarity, we should distinguish 'movements' and 'activities', such as the journey-to-work or the production of a good, which may sometimes be called 'processes' by some authors, from those processes which change, or aim to change, the underlying pattern. We try to restrict our use of 'process' to this latter category.)

In Chapter 2, we identified the main entities which make up cities and regions, and in this chapter, we have begun the process of describing the main geographical features of these systems which form the subject matter of theory. We have emphasised the focus on the *location* of activities and associated infrastructure (whether of people or of organisations), noting that we can either look at the distribution of activities across a region or at the mix of land use within each zone. We have also emphasised the range of *interactions* between activities at different locations. The existence of this degree of interdependence, which is accentuated by the competition for land and other resources, provides the interest and difficulty in many theoretical problems in urban and regional analysis.

At any one time, there is a pattern formed by the existing distribution of variables across space. Out of these distributions, certain aggregate quantities can be calculated, such as densities and accessibilities, which form part of the basis for ongoing decisions that will generate at least marginal changes, and sometimes greater changes, in the system as a whole. To understand and theorise about these *processes* of change, it is necessary to understand the behavioural basis provided by the various agents of the system. Thus, there are two types of theoretical question: one involved with explaining the overall pattern at a particular point in time; and a second with explaining the marginal rates of change at any time (or any non-marginal changes if the system is not stable). Of course, in an ideal theory, the two would come together; the second and dynamic theory would be run over a long period so that it also predicted structure. In practice, data difficulties, if not unsolved theoretical problems, prevent this from happening, and so it is useful to recognise that either of the two approaches may in principle contribute to theory in different circumstances.

Finally, we need to return to the points which have been stressed about the consequences of interdependence. Because of the high levels of connectivity between different entities and subsystems, there is a strong argument for building a very comprehensive model within which these could always be taken into account. However, once again, the ideal solution is very often impractical and we have to recognise that partial approaches, for all their imperfections, may still be useful. These alternatives (which also interact with the 'level of resolution' issue) can be borne in mind as we proceed.

In the same spirit, it is also worth remarking that we have identified some very general features which are the main elements of theory: location and land use mix, interaction and network flow, and the way in which these variables are determined at various scales by the optimising decisions of agents. This suggests that it is also worthwhile to attempt a general approach to theory. The impossibility of always carrying this off in this case may lead us to develop theory which is only applicable in a particular sector.

Again, we will always be seeking a balance of interests between the ideal and the practical.

In the next chapter, as a final preliminary on theory, we examine the potential contributions from a range of disciplines. First, however, we consider the third 'dimension': method.

Methods

INTRODUCTION

We take this part of the argument in two stages. First, we discuss the technical methods available to the urban and regional analyst for the representation of understanding and for turning theories into formal models. Secondly, we consider the methodological demands of planning and problem-solving with models. We also need to recognise a broader arena: we need to discuss the status and validity of methods – the philosophical underpinnings and critiques of different approaches. There are a variety of approaches to knowledge: what we learn from a novel about the people of a city may be more valuable than a mathematical model. It depends to various degrees on the purpose of a particular inquiry. This second arena therefore raises the broad issues of the methodological offerings to urban and regional analysts – first from philosophy and then from a variety of disciplines which share interests in urban and regional systems. In this section, we focus on technical aspects of geographical method and we return to the quasi-philosophical issues in Chapter 4.

THE TOOL-KIT

Introduction

A wide range of methods, or tools, are available to help the analyst. These methods can be worked on and understood at different levels and in different ways. This book interacts with a number of others in this respect. The methods are often mathematical ones, though it should be emphasised that the main ideas of geographical theory can often be understood with the use of only elementary mathematics. Such a background is provided by the first three chapters of *Mathematics for geographers and planners* (Wilson and Kirkby, 1980), and the rest of that book also provides much useful material. That book is essentially organised as a mathematics text with geographical illustrations. An alternative perspective is provided by *Geography and the environment: systems analytical approaches* (Wilson, 1981a) in which the discussion of methods is organised around the types of system used by the geographer. *Mathematical methods in human geography* (Wilson and Bennett, 1985) is a guide to the more detailed mathematics that might be needed. A recent account of model applications, including presentations of the relevant mathematics, can be found in *Intelligent GIS* (Birkin *et al.*, 1996). All this previously published information enables us in *this* book to focus on the main ideas of the various theories without getting into very detailed mathematics. Finally, it should be mentioned that there are other associated books within which aspects of geographical method are treated in much more detail. They will be cited as appropriate in the text, but the particular offerings of the present author (and colleagues) which couple directly with this scheme cover entropy maximising (Wilson, 1967, 1970), population accounting methods (Rees and Wilson, 1977), optimisation methods (Wilson *et al.*, 1981) and catastrophe and bifurcation theory, which provides much of the basis of dynamical modelling (Wilson, 1981b).

A technical piece of analysis can be considered to progress through four stages (not all of which will necessarily be present or visible in any one exercise). They are as follows:

(i) preliminary systems analysis and theorising;
(ii) statistical analysis;
(iii) mathematical modelling and formal systems analysis; and
(iv) applications in management or planning.

The first stage would consist of the first run through the first five steps discussed in this chapter: entitation, scale, spatial representation, partial–comprehensive and preliminary ideas about theory, including an appropriate conceptualisation. This would give quite a detailed picture of the system of interest and associated analytical problems (which might be modified in the light of subsequent research). The mention of systems analysis in this preliminary stage is intended to reflect a concern with interdependence and the study of underpinning relational structures.

The results of the first preliminary stage may well be interesting in their own right. It could be argued, for example, that much 'traditional' geography was of this form. It can also be argued that in the application of the ideas of modern quantitative geography, this stage has been skimped or omitted and some of the results have consisted of rather simple-minded applications of quantitative methods.

The next two stages – statistical or mathematical approaches – are distinguished by whether an inductive or a deductive approach is preferred. In the end, this distinction becomes rather blurred: some pre-analytical theoretical hypotheses – in the selection and categorisation of variables, for example – are required by statisticians; and goodness-of-fit tests and calibration procedures are needed by mathematical modellers. The choice is likely to be made in relation to the complexity of the system and associated theory and whether appropriate mathematical methods are available or not. When not, the simpler statistical techniques form an (approximate) fall-back position. Perhaps choice of method is most often determined in practice by the skills and training of the particular analyst.

We proceed now, therefore, by describing briefly first the main techniques that are available from statistical analysis and then the greater variety of methods of mathematical systems analysis. Details are available from the books cited earlier and some of the main ideas will be sketched in relation to examples later.

Statistical analysis

We can summarise the techniques available, in roughly increasing order of difficulty and complexity (and usually data requirements!):

(i) *The description of geographical entities.* This includes the use of the standard distributions and their properties and simple hypothesis testing of straightforward relationships.

(ii) *Use of standard statistical models.* This includes the use of the general linear model to investigate geographical relationships and the elementary modelling of trend surfaces, time series and spatial dependence. It is also possible now to develop nonlinear models.

(iii) *Models of spatial relationships.* Attempts can be made to 'explain' a static picture using the models of (ii) above, but it is now more common to investigate spatial structures as arising out of spatial processes, and geographers have developed their own techniques for this purpose.

(iv) *Models of temporal relationships.* Similarly, more modern methods can be used to explore the relationships between variables over time, including the task of modelling changing parameters.

(v) *Spatio-temporal relationships.* The obvious target of the statistical geographer is to model spatial processes, structure and temporal change simultaneously. Hagerstrand's (1953) diffusion model was an early attempt at modelling spatial dynamics.

Mathematical modelling and systems analysis

Here, we follow the argument of Wilson (1981a) that when systems analytical methods are applied in geography, it turns out that a number of mathematical modelling methods can be specifically tailored for geographical purposes. They can be summarised as follows:

(i) *Algebra and analysis.* The basic tools of algebra and calculus allow us to express many of the functional relationships between the variables which describe geographical systems. It turns out to be very important to find the most efficient ways of proceeding in terms of algebraic description, particularly with respect to choice of spatial representation.

(ii) *Entropy-maximising methods.* These play a particular role in a number of models to be presented in Chapter 6 of the book. They are essentially concerned with finding the most probable state of a system and represent one of the general ways of building theories for Weaver's systems of disorganised complexity.

(iii) *Account-based methods.* The interdependencies between elements of geographical systems often force a number of accounting relations to hold between them. These are often valuable components of theories. In some cases, again for systems of disorganised complexity, they can also provide the basis of another kind of statistical averaging method which can be used in theory-building, especially at macro and meso scales.

(iv) *Optimisation methods.* Many of the processes underpinning theory are concerned with maximisation or minimisation. The mathematical tools for these situations have advanced very rapidly in recent years, providing tools for the theoretical geographer that allow enormous strides to be made relative to the work, for example, of the classical theorists (see Wilson *et al.*, 1981).

(v) *Network analysis.* A particular feature of geographical systems is that flows of different kinds are carried on networks, and these are important components of many models. It is valuable, therefore, again to have a set of specific mathematical tools available to help in this aspect of theory-building (see Haggett and Chorley, 1969; Scott, 1971).

(vi) *Dynamical systems analysis.* This is now at the heart of complexity theory and will be developed in more detail in Chapter 6.

APPLICATIONS IN MANAGEMENT AND PLANNING

Since geographical analysis relates to systems for which many individuals, organisations and government agencies have concerns and ambitions, the results of the analysis ought to be valuable for the management and planning of such systems. This means that it is necessary to understand the methods of management and planning both in general and specific contexts and to be able to marry them, where appropriate, to the methods of geographical analysis.

It was argued many years ago in a framework which has stood the test of time (e.g. Wilson, 1974, chapter 2) that planning (and we can, for convenience, take management as 'short-term planning') can be divided into three kinds of activities: policy, design and analysis. The first of these, *policy*, is concerned with the specification of goals and the evaluation and choice of and from a number of alternatives (one of which will always be 'do nothing'). The second, *design*, has the task of generating alternative plans. The third is concerned with systems *analysis*, predicting future problems and needs, and predicting the impact of plans and hence the material which forms the basis for evaluation procedures. Analysis is, therefore, very much what we have taken it to be already, as geographical analysis. If our range of methods has to be extended, therefore, it is in terms of design and policy.

KEY IDEA 3.1

The ways of thinking associated with policy, design and analysis are different. They need to be understood and nurtured. Although the focus of this book is analysis, the broader context should be borne in mind.

A plan consists of settings of quantities that are controllable – the number of houses permitted in each zone of a city for example. The design of plans is a matter of invention, and then testing through an analysis of impacts. In some cases, however, it is possible to use optimisation methods, such as mathematical programming, to attempt to generate an optimal plan. In the case of policy evolution, the main methodological problems are those of the measurement of costs and benefits and we return to these in the discussion on economics in Chapter 4.

This kind of approach to planning as policy, design and analysis implies a well-defined governmental and planning structure. It is also worth remarking, as will become more clear in our discussion of the social sciences in Chapter 4, that these government and planning structures are themselves matters for analysis.

The procedure to be used in this book is that the bare essence of the appropriate methods will be described in the context of each example as it is presented. It will then be left to the reader to pursue the ideas in more depth in the books already cited.

Summary and conclusions

We need to define systems of interest of some coherence. That is, we put together those entities which are strongly related to each other. It is also clear that we will assemble systems of interest at a variety of scales – ideally with the different analyses, from coarse to fine, explicitly and hierarchically related to each other. We would normally attempt to use as fine a spatial representation as possible, but there will be limits to this imposed by the availability of suitable methods of analysis for handling very large numbers of variables and also by data availability. On the partial–comprehensive spectrum, we obviously need to be as comprehensive as possible, though again practicalities often inhibit this. In the rest of this section we discuss in outline the interplay of these choices in different kinds of situations and we investigate the kinds of theoretical and methodological problems that are thrown up by different kinds of choices.

There is some possibility of trade-off, for example, in relation to scale and comprehensiveness. Given that comprehensive theories are more difficult to build, one way of making progress has often been to attempt comprehensiveness at the cost of working at a more coarse scale. Central place theory provides a good example of this: large and complex settlements are treated as points without any internal structure of their own. This kind of approximation sometimes works at least to the extent of offering some insight, but it should be emphasised that interdependence of the type discussed in the previous paragraph is then lost. The question for the theorist is then: is the approximation adequate for the uses to which this particular theory is being put? Our objective, therefore, is a comprehensive approach at a sufficiently fine level of resolution for all the interesting interdependencies to be built into the theory.

One further condition can then be added in relation to theory development: that the formulation should be capable of explaining the development of the system of interest through time. This involves an explicit treatment of the processes which generate change. Again, this is an ideal. A first approximation may be to build a theory which 'explains' a static cross-section; but this will usually be found to be inadequate in the end.

We can conclude this chapter by reviewing the interplay of the theorist's decisions on specification of system of interest, scale, whether to be partial or comprehensive, spatial representation and structure and process. This is a high-dimensional space which we now sample to provide a concluding overview.

At the micro (and indeed other) scales, we group together some of the elements described in the first section of the chapter. We distinguish households, farms, manufacturing firms, providers of services (in each of these cases noting that they can be in the private or public sectors), and plots of land. The partial–comprehensive dimension generates a continuum of problems. The simplest, in each case, involves taking one unit and assuming that the rest of the system is given. In such cases, the units would usually be represented as points, with the elements of the environment represented as points, zones or continuous space (through, in this last case, something like density functions). If points, they could be allowed to 'move' in continuous space, or be members of a fixed discrete set, perhaps located on a network. A theory of pattern would then predict the location of the unit according to some basic principles, like maximisation of utility, profit or public welfare (or others) as appropriate. It would also predict the spatial connections of the unit to other locations: its spatial interaction pattern. A dynamic theory would focus also on potential and actual change, either in level of activity (at the same location), birth (a new unit entering the system), death, or migration (change of location).

As the treatment becomes more comprehensive, the theory deals with more units. In the first instance this is likely to involve only units from the same sector, still taking everything else as given. These may be in competition or be interacting in some way, and this interdependence will at least in part determine the locational pattern. As the degree of comprehensiveness increases, units in other sectors will also be made endogenous to the theory. For example, the location of manufacturing jobs may be taken as fixed and part of the 'environment' but both household and retail service units may have locations determined within the theory.

At the meso scale, the same sectors can be distinguished, but we also add 'settlements' which can be visualised as entities at this scale, and transport and communications which, as observed flows on networks, are essentially meso scale phenomena – aggregates resulting from the adding together of larger numbers of individual decisions. The residential, agricultural, industrial and service sectors will now also be treated as aggregates in a similar way. This has an impact on the spatial representation. It is possible to treat space continuously or in discrete units – zones which are mutually exclusive but which cover the region in an exhaustive way. The level of activity of a sector in a zone, say of a manufacturing industry, or retailing, may then be treated as though it was rather like a firm at a micro scale even though typically it is made up of a number of firms. The output of theories at this scale will be a spatial distribution of activities by sector and a representation of transport and communications flows.

The partial–comprehensive dimension works in a similar manner to that at the micro scale: in a partial treatment, a theory will be concerned with one sector, taking all the others as providing a given environment; as the approach becomes increasingly comprehensive, more sectors will be allowed to interact. The dynamic theory will be more concerned with 'level of activity' changes rather than a more direct representation of birth, death and migration processes because of the loss of resolution. The outputs of theory at this scale will include a number of measures which are basic meso properties of locations: these include densities and intensity-of-use measures, accessibility measures and congestion. It is through these variables that theories interact at different scales because they can play a part, as locational properties, in the decision processes at the micro scale. Another example of a meso scale property is 'hierarchical structure': this is determined by the mix of activities which make up something like a settlement, and this only has meaning at this scale.

At the macro scale, for convenience we only distinguish two sectors, though there could obviously be much more sectoral disaggregation. This scale provides a summary, for a whole region, say, of the make up of the population and the economy. As we will see in Chapter 4, much demographic and economic theory is aggregative (macro) in this way. It provides an important backcloth for geographical theory at the other scales: population totals, which form the basis, at least in part, of the demand for goods and services; and the nature of the economy, which determines the possibilities of supply of goods and services. And macro economic theory can help to determine some quantities (many prices, for example) at this scale, which are then directly usable in micro or meso scale models.

In each of the problem areas that have been defined, the theoretician will have to seek to understand the basic processes at work. Many of the geographical problems, across sectors, have many common features but are then distinguished by the different objectives of the units in those sectors. We will see examples of the consequences of these differences as our account of the development of theory progresses.

4

Conceptual foundations: urban and regional analysis as a multidisciplinary programme

Multidisciplinarity

We have already set out in some detail an account of the components of a city or a region. The key disciplines that can contribute to understanding will clearly fall in the social sciences – economics, geography, sociology and psychology provide examples. Life in cities, the record of experience over the centuries, will be described by historians and novelists as much as by social scientists. And as for most disciplines, the methodological disciplines also provide some of the foundations. There are also professional disciplines: urban and regional analysis will be connected to the applied sciences through the *engineering* and *architecture* of cities and regions; the operational and policy aspects will be reflected in fields such as city and regional planning and social administration. It only takes this brief sketch, therefore, to show that urban and regional analysis is *essentially* multidisciplinary.

Fortunately, the approach in this book is highly circumscribed. By adopting a *modelling* focus, the objective is to provide the accounting and mathematical basis of our understanding. This approach both draws on the full range of other disciplines for its concepts and aims to provide a framework within which other disciplines can operate – by saying something about the size, shape and dynamics of cities and regions, and about their populations, organisations and infrastructure.

What do we need from this review to help us establish the foundations for urban and regional models? We can use the STM framework of the previous chapter to answer this question. We will need to identify the best kind of *systems articulation* and representation for modelling purposes – though this framework will also enable us to cast a critical glance at the representations of cities and regions used in particular disciplines. For *theory-building* purposes, we need a tool-kit of *concepts* to facilitate understanding, to use in model-building – and to connect what can be achieved through modelling to what is achieved in distinct disciplines. For *methodological* contributions to modelling, we need a tool-kit of a rather different kind which will enable us to operationalise our theories.

We begin by looking at the substantive disciplines in our search for concepts. The first of these is concerned with demography – an obvious starting point, since we must at least be capable of counting the populations of cities and regions. Then we turn to economics as the social science that is most likely to give us the core theory of the behaviour of individuals and firms and of economies in aggregate. Location has not been a key feature of economic analysis, but it is, of course, for geography, which is also covered. We then collect together the remaining social and humanities' disciplines.

We then review the methodological disciplines. We begin with philosophy, partly because modelling attracts controversy in relation to its position in the social sciences in general and in urban and regional analysis in particular. It helps, therefore, to position it at the outset. We cover the mathematical tools; we review what the sciences have to offer; and it is here that we locate the composite 'disciplines' of systems analysis and complexity theory. In the following section, we review the professional disciplines – engineering, architecture and planning – and note the common element of *design*.

In the end, we will want to recommend an integrated approach for urban and regional analysis, and we have already noted that modelling provides the framework for this. It is not surprising, therefore, that those who have been drawn into urban and regional modelling from their own original discipline have in many cases attempted to incorporate the whole package within that discipline. We review some of these activities in the final section, and examine the new disciplines, such as *urban studies* or *regional science* which have been established to fill the gap.

The substantive disciplines

DEMOGRAPHY

Demography is concerned with explaining the sizes and compositions of populations. Since it is concerned with counting, it is, not surprisingly, a model-based discipline. It has a long history, essentially dating back to Malthus. He produced a model which represented exponential growth and it was only much later that this was transformed into the more conventional logistic growth model. We review these developments in Chapter 6 in the context of the broader history of modelling.

Much of the work of demographers has been incorporated directly into urban and regional analysis, e.g. through the subdiscipline of population geography. Indeed, the geographer's concern with space has led to an emphasis on aspects of migration which can reasonably be fed back into demography. This has provided the basis for the development of multi-regional demography though the developments which enabled this came, as we will see, mainly from outside the discipline

itself (Rogers, 1973, 1975; Rees and Wilson, 1977).

It is worth noting at this stage that demography valuably demonstrates a general modelling principle: that of *accounting*. The notion that it is possible to account for the life cycle of each individual member of a population both builds a certain kind of consistency into models and has, as it has turned out, provided a number of more general modelling techniques (cf. Stone, 1967, 1970). We develop this as a key idea in Chapter 6.

ECONOMICS

Introduction

Economic theory is concerned with the behaviour of a variety of agents – producers, consumers and public bodies, for example – and the relationships between them. It should, therefore, be a good source of component hypotheses for the urban and regional analyst. Indeed, it has generated the subdisciplines of urban economics and regional economics which encapsulate these possibilities from within the discipline. However, these have not been the most prestigious branches of economics. An American colleague once observed that 'putting the word "urban" before "economics" has the same effect as putting the word "horse" before "doctor"!'. Nevertheless, the practitioners in these new subdisciplines have generated very high quality research – sometimes constrained by the typical STM decisions of their economic peer group.

The joint actions of the various agents determine the pattern of goods and services that are produced and consumed and the prices that are charged for them. In this section, we focus on some of the concepts and principles that have been introduced by economists and which will be the building blocks of a number of theories in urban and regional analysis. Only the broadest sketch is offered at this stage.

Basic principles: the neoclassical position

Classical economics was very much concerned with the nature of production and particularly costs. Relative prices, for example in the Ricardian system, were assumed to be determined by the relative costs of inputs, and ultimately, all these costs derived from labour inputs. Hence prices were determined by a labour theory of value. This scheme was amended by the neoclassical economists by an expanded treatment of the demand side: preferences were recorded in utility functions of some kind, and the maximisation of utility was the basis for the derivation of demand curves. It is then the intersection of supply and demand curves, rather than simple factor inputs which determine prices. We concern ourselves here mainly with this neoclassical theory, though later we add a note on the main alternative formulation which is based on the work of Marx.

As might be expected from our earlier discussions of the elements of urban and regional theory, there is a major aggregation problem in economic theory. At the micro scale, the behaviour of agents is modelled usually as though they can have no influence on certain macro quantities like the determination of prices of goods. At the macro scale, theories then have to be developed to explain these quantities and relationships. It then becomes a separate theoretical problem to relate the models at the two scales. (This is another version of the agency–structure problem.) In economics there was for a long time relatively little work at what we have called the meso scale at which spatial issues were clearly represented. This changed with the rapid development of urban and regional economics (see Nijkamp, 1986, for example). In the main, we will draw on this work as 'geographical' theory in later chapters – 'geographical' in the sense of being concerned with space; no interdisciplinary competition for territory is implied.

At the macro scale, we will note only the simplest of the principles involved. It can be assumed for goods which are sold in 'markets' that the quantity produced and the price is determined by the intersection of the well-known demand and supply curves for that good. The underlying assumptions for this derivation include the existence of a perfect competitive market and that a stable equilibrium can be achieved. For goods and services which are supplied by public agencies and for which there is no market, then this can be considered, at its simplest, to be achieved on a cost minimisation basis, or by the maximisation of a social welfare function if such a thing can be constructed and measured. In either case, the theory has to be related to what is assumed to be happening at the micro level and it is to this that we now turn.

The analysis of production is based on the theory of the firm. Each firm is assumed to maximise profits. It is important also to focus on the production function of the firm, which describes what the technical possibilities are for different combinations of factor inputs: capital, labour, land and other goods. If the prices of all factor inputs are known, and firm itself cannot affect them, and the prices of the outputs are known, then it is a straightforward maximisation problem for the firm to choose the mix of inputs and level of outputs that maximises its profit.

The theory of consumer behaviour is based on the notion of the utility function. This is a function of variables representing the quantities of goods and services purchased by each consumer and is a measure of the value (in some appropriate units, not necessarily money) associated with any combination of purchases. In effect, a consumer has to be able to rank alternative combinations for a utility function to exist. This function is then maximised subject to constraints on, for example, income. (In some cases this may be made exogenous, since part of the choice is how much labour to sell in order to generate an income.) Preferences, and hence utility functions, can, of course, be assumed to vary across the population.

The two theories are similar in character: a function representing the desirability of the outcome of choosing from a set of alternatives is maximised subject to constraints. They will provide the basic underpinnings of the models to be presented in Chapters 5 and 6, though within a different framework which is both more practical, and hence more powerful, analytically; and which represents market imperfections.

KEY IDEA 4.1

Utility and profit maximisation (and equivalent public sector measures) can be used as a basis for modelling provided market imperfections are explicitly incorporated.

As we will see in other contexts in later chapters, the specification of the constraints is a difficult task: they represent the complex 'environment' faced by any decision-maker and are an important part of theory-building.

The workings of the whole system are immensely complicated. Producers are competing for factor inputs (and hence help to determine their price) as well as for markets for their goods. There is also the important associated concept of *rent*. The owners of factor inputs such as land or raw materials can obtain payments arising from the fact of their ownership, and the level of these payments will be determined by their most profitable use: this fixes the rent per unit of such inputs. This notion is obviously particularly important in theories about land which is so important in urban and regional analysis.

It is assumed that an equilibrium can be achieved for all the interactions in the various markets, and the results of this at the macro scale are the quantity–price relationships referred to earlier. In practice, of course, the world is much more complicated. There are imperfections of many kinds in markets and much work has been undertaken by economic theorists to describe and explain these. One of the theories of the role of the state in economic management is that it exists to rectify imperfections of this kind.

The distribution of income is in part determined by the distribution of ownership of factor inputs and the means of production (capitalists) and in part by the market in labour through which people sell their time and skills. One of the features of neoclassical economic theory is that it does not make any statements about desirable distributions of income. There is an implicit assumption that the distribution which is generated by the market is the correct one. In practice, this is not always borne out and another role of government is considered to be the management of the distribution of income through the tax system and the provision of other benefits.

Many commentators draw attention to the importance of the concept of real income in this context, not simply money (wage, salary, profit or rent) income. This will include an amount (which is often difficult or impossible to measure in money terms) representing the benefits of public services that are received. In recent years, welfare economists have devoted much effort to attempts to measure 'social' benefits, and such measures can then be used to improve the allocation of resources in the public sector. In contemporary analysis, there will be particular concerns with environmental benefits and the extent to which they are valued in money terms.

The dynamics of this system are governed by the size of the economy, its external environment and by the various functions which have been introduced above. An increase in population, for example, will lead to increased demand for goods and services and to an expansion. Changes in the volume or terms of trade will affect the scale of production of goods from import and export. Technological change will modify the production functions of firms and lead to the creation of new firms and others becoming obsolescent. It will change the mix of factor inputs; at present, for example, this is represented by a tendency to replace labour by capital. Changes in public taste will modify utility functions and hence demand. Political changes may lead to modifications in the distribution of income. The impacts of any one of these changes will have repercussions throughout the system because of the interdependence of the different elements.

The Marxian alternative

The alternative tradition in economic theory has been that generated by Marx. He devotes greater attention to the mode of production and in particular argues that the capitalist mode of production creates social relations between the owners of capital and labour which form the main dynamic of the economic system. Less attention is given to individual utility and the exercise of choice. Profits are seen to be derived from the surplus value added to production processes by labour and this surplus is appropriated by capital. The price of labour is fixed at its minimum level subject to whatever improvements can be made through class struggle. This also leads to a prediction of a falling rate of profit in the long run, and a dynamic which assumes, because of the internal contradictions within the capitalist mode of production, that the system will eventually collapse and be replaced by a socialist one.

From the point of view of theory as treated in this book, perhaps the most important lesson to learn from this approach is that it is essential to look in detail at the way different agents in an economic system operate. It provides an emphasis to the point that the neoclassical theoretical system is a model which does not always describe in any detail what is actually happening; and in some cases may not describe reality in principle either. However, it is not clear, certainly at the meso scale, that a new detailed basis for urban or regional theory is yet available. It could well be that, if appropriate emphasis is given to constraints and the ways in which these are built into theories, the neoclassical and Marxist approaches could move closer together.

Economics in urban and regional analysis

It is useful to conclude this section with an outline of a number of 'principles' derivable from economic theory which Paelink and Nijkamp (1975) argue form the basis of the contribution of economics to urban and regional problems, and then to discuss briefly, again following their argument, a number of special features of this field of application. These form a useful summary of much of the earlier discussion. Paelink and Nijkamp draw attention to the following:

(i) *The activity principle.* In effect this focuses on production or consumption processes, noting that each is defined by its mix of inputs per unit production of outputs, so that the total production is determined by the level of activity that is adopted. This is equivalent to focusing on production or utility functions and the appropriate constraints, but in relation to the technical possibilities.

(ii) *The substitution principle.* This states that a given level of output, of goods or of utility, can be achieved from different combinations of inputs, and one of the tasks of economists is to chart these substitution possibilities.

(iii) *The optimising principle.* The actual choices which are made then arise from optimising principles: maximising profit, minimising cost, maximising utility, and so on.

(iv) *The dispersion principle.* Because markets do not operate perfectly, outcomes are not the result of optimising principles. So a way has to be found for theories to incorporate this fact while still being determined by underlying optimising behaviour.

At a later stage of their argument, Paelink and Nijkamp discuss some features of economic systems which are of particular importance in urban and regional economics:

(i) The existence of *indivisibilities* is important in many production processes, and these lead to *economies of scale*. It is this feature which leads to the spatial concentration of many activities. These are sometimes called *agglomeration* economies and are of particular importance in urban and regional analysis.

(ii) There are often *external economies* (or diseconomies) arising from production processes. Transport costs for factor inputs might be reduced if a lot of firms are in spatial agglomerations. In the same circumstances, other firms and the local population may have to bear the pollution costs of a particular process.

(iii) The structure of the spatial economy will be partly determined by the *transport services* which are provided at various costs.

We will deploy many of the concepts introduced above in a variety of ways in subsequent chapters and elaborate on them as need be in those more specific contexts.

How much progress have economists made in capturing their own theories about cities and regions in models? Three preliminary comments can be made. First, it is clear that different economic sectors will be interdependent in macro-economies and the major advance which represented this was made by Leontief in his development of the input–output model (see Leontief (1967) for a summary). As in the case of demography, this was significant not only for economics, but as a modelling technique which has much wider applicability and we will pursue it as such below at various stages in the argument. Secondly, economists have developed models of cities and regions which represent spatial structure.

Richardson (1977) still offers one of the best summaries of these contributions. Relevant collections of papers are offered by Nijkamp (1986) and Mills (1987). We discuss these in the wider context of urban and regional analysis, first as they have been deployed as key components of classical geographical theory in Chapter 5 and then in terms of the history of modelling in Chapter 6. Thirdly, when models are applied in urban and regional *planning,* we will be concerned with evaluation methods, and economic analysis has a crucial role to play in many aspects of the measurement of value. This has been encapsulated in the development of cost–benefit analysis and the appropriate concepts will be introduced below. It has been particularly important to tackle issues such as the value of time in modelling, and the original drive for this came through cost–benefit analysis (see Foster and Beesley, 1963).

In recent years, there has been a resurgence of interest of economists in urban and regional issues led by Arthur (1988, 1990, 1994a, and see Arthur *et al.,* 1997) and Krugman (1993, 1995, 1996) but much of this work has been in the context of complexity theory; this is taken up later in this chapter as well as in Chapter 8. There have also been recent reviews in the two volumes edited by Arnott (1996a, b) and Anas (1996).

GEOGRAPHY

In Chapters 2 and 3, we have already noted that cities and regions are prime systems of interest for the human geographer. Geography necessarily therefore functions as an integrative discipline – but it has not been fully successful in this respect. It has absorbed the relevant concepts from other disciplines but has deployed these in a variety of subdisciplines ranging from the primarily descriptive to models which are rooted in what might be called classical theory. The power of these classical approaches within the discipline is so important that Chapter 5 and Appendix 2 are devoted to a full exposition. One of the central concerns of the book is then the argument that the full power of the new modelling can be used to rewrite this theory in modelling terms, and this is tackled in Chapter 7 (following the exposition of the key ideas in Chapter 6).

To provide the context, we briefly review the relevant aspects of the development of human geography as a discipline. There has been a progression from traditional *regional* geography to *systematic* geography – in effect, different articulations of the systems of interest; and then developments through new theory and methods. The first phase of development was through statistical analysis and formal approaches to patterns (cf. Haggett, 1965; Haggett *et al.,* 1977, for the classic texts) and, separately, modelling, which facilitated the handling of complexity. A second phase was through *radical* or *structuralist* approaches with a bigger input of contemporary social theory to balance the economic dominance of earlier styles. These different streams of interest all continue to develop, but inadequate communication between them is evident. Regional geography will continue to remain potentially valuable as a synthesis of knowledge and as a basis for studying local problems and for planning purposes. It should have been immensely strengthened by developments in systematic geography, spatial analysis and structuralist geography – and indeed to an extent it has been, but not enough. Regional geography has come to be identified with 'area studies' and the practitioners are typically not able to make use of many of the techniques available to them. Similar comments could be made about the relationship of systematic geography to spatial analysis. Texts are still written on, say, economic or urban geography which make no use of advances in location theory or modelling. Further, uncritical applications of models are carried out with no reference to the questions raised by radical geographers. Structuralist geography, on the whole, has fared better in its impact on systematic geography and is now starting to have an impact on what might still be called regional geography through 'locality' studies. To take the wheel full circle, some social geographical theory cries out for modelling techniques as an aid; but they are little used.

The position adopted here is that an understanding of modelling as a framework for theory development provides the foundations for other theoretical positions. The methods used for model-building are not strongly dependent on their theoretical components, for example. It is relatively straightforward to substitute alternatives (e.g. Marxian for neoclassical) if desired.

THE REMAINING SOCIAL SCIENCES, HISTORY AND LITERATURE

Social scientists share the aims of the human geographer in a broad sense, but because they have shown relatively little interest in spatial analysis, or in synthesis with respect to place, the development paths of the two groups of subjects have been very different. In this section, we try to outline the concerns of the social scientist and aspects of the works of others in related fields, though we make relatively little use of this work in the next two chapters of the book which contain the descriptions of the core of geographical theory.

The most obvious subjects to group together in the first instance are sociology, anthropology, politics, social psychology and psychology. They are all concerned in various ways in seeking explanations of the behaviour of individuals and organisations. Such studies begin with seeking productive forms of classification. There are also obvious demarcation lines created by scale: the psychologist is unremittingly micro, taking the social environment of subjects as given. The other subjects tend to operate on a micro to macro continuum even though the ultimate goal may be to seek explanations of social structure and change, i.e. at macro or meso scales. A fundamental question, analogous to the aggregation problem in economics, is the previously noted 'agency–structure' problem: how do individual actions help to create structures while simultaneously being constrained by those structures?

At the most basic level, the social scientist is concerned with identifying different kinds of agents. This leads to notions of class, one example of which we have touched on briefly in citing Marxian economics above. What differentiates much work in the social sciences from corresponding work in geography (though it should be said immediately that the distinction is by no means a sharp one) is a concern with the underlying motivations and values associated with behaviour and practices at different scales. Some of this emerged as a response to what has been seen as

'positivistic' approaches – attempts to develop 'value-free' social science along the lines of the natural sciences. Not only do values need to be identified, but also their sources, and the connections of these to power structures. This leads in turn to a search for underlying ideologies in different societies and an investigation of the 'consciousness' of different members of those societies. This can be important, of course, in relation to the sorts of economic theories outlined in the previous section. It can be argued that neoclassical economics is based on the ideology of the capitalist class – that it is not surprising, for example, that it has no lively concern with the distribution of income, since it is part of that ideology to believe that the distribution of income created by the market is 'right' and 'good'.

These questions are shared by those whose (social scientific) profession is politics, but typically with more emphasis on the role of the state in society. There are a number of theories about the nature of the state and why it comes about in various forms. There is, however, no single theory yet available that we can use as a contribution to geographical theory, though again this raises questions, when we look at applied geography and planning, to which we shall have to return later.

Two other disciplines have been drawn into the argument in this section: history and literature. This is because they are both concerned, at least in part, with the same questions as the social scientist: with the understanding and explanation of individual and social behaviour and the nature of the corresponding societal structures. Giddens (1979), indeed, has argued that there is no distinction between sociology and history. We would broadly agree, but it is appropriate to draw attention to one feature of the practice of history as a discipline from which important lessons are to be learned by others: that is, the notion of 'evidence'.

The historian, in trying to put together an analysis to 'explain' a sequence of events, faces all the questions of the social scientist. Typically, however, the questions relate to longer periods of time and frequently to much more distant periods of time. The second aspect means that, while the social scientist might in principle try to assemble all the data needed to test a theory, the historian can rarely do so, and so concentrates on assembling relevant evidence. It is appropriate for other social scientists, including human geographers, to recognise that, whatever their ambitions, they are most often in the same situation. It is like trying to solve a jigsaw puzzle where only a fraction of the pieces are available, and indeed in many cases where the pieces that are available have come from different manufacturers and are of different sizes so that adjacent pieces do not fit together properly.

Writers, novelists in particular it can be argued, have similar aims. In creating 'believable' characters, the novelist has to incorporate what amounts to an effective theory of how the world he or she is describing works. At this point we simply note that much 'evidence' – even though it is called 'fiction' – is available to us from this source. It can even be quite direct: J. G. Farrell (1970), for example, in his novel *The Singapore grip*, includes detailed descriptions of the workings of the rubber industry in Singapore and Malaya in the 1930s and offers more insight than would be obtained from most economic texts. Hudson (1972) provides an excellent statement of the general argument that the novelist can offer much more than the social scientist, especially in relation to his own subject of psychology.

Since part of the focus of the book is complexity theory, it is useful to note two recent publications which summarise the potential connections between complexity theory and the social science and humanities' disciplines. Kiel and Elliott (1997) review 'chaos' in the social sciences and there is one contribution, that by Dendrinos (1997), which bears directly on urban and regional analysis. What is particularly interesting, as a precursor to the next section, is the essay by Harvey and Reed (1997) – building on the work of Boulding (1968) and Smelser (1963) – which matches styles of modelling against what is, in effect, the depth of the social science questions being posed.

In the humanities, Hayles' (1991) edited collection includes essays which apply the ideas of complex dynamics in literature, including one essay, that by Porush (1991), which focuses in part on traffic! However, his approach is through Prigogine's work rather than the interaction models used here.

Methodological disciplines

PHILOSOPHY

Introduction

A methodological discipline is defined as one which does not have substantive systems of interest on its own account. As ever, this is an approximation, but a useful one. At the most general level, we should begin with *philosophy*. Its potential contributions to theory in urban and regional analysis come from at least two distinct directions. The first arises from the theories of knowledge and meaning – or the closely related concepts used in informally introducing the notion of theory in Chapter 3, truth and understanding. A wide range of concepts are used in urban and regional analysis and it is helpful to be aware of the methods for the scrutiny of these that are available. This is particularly important at the most general level, as we hinted in another context in Chapter 3, with such concepts as 'theory', 'hypothesis' or 'law' in relation to the notion of 'truth'. It is perhaps crucial when many social scientists, perhaps geographers in particular, are brought up in a tradition that regards a subject like geography as concerned with 'facts'. When the discipline of philosophy is brought to bear on the observations that are presented as 'fact', then a more critical assessment can be generated. There may be problems of perception and measurement, for example, not to mention selection (of data). Critical scrutiny of this kind can be helpful rather than, as might appear at first sight, crippling, because it allows pragmatic solutions to be adopted provided the authors are well aware of what they are doing. What we perhaps look to philosophy for most of all is a push towards clarity in our use of concepts.

We can also learn from the theory of knowledge of the important distinction between inductive and deductive theorising. Induction consists of the attempt to infer 'truth' in the form of 'laws' from data, while the deductive method consists of the invention of theory, the predictions of which can then be tested against observation. If appropriate, the theory can then be refined. 'Truth' becomes related to something like 'degree of belief' in a theory: we can say that a theory is 'true' in the colloquial sense if it has been successfully tested frequently. Typically, it would be necessary to add riders noting any restrictions on the range of conditions under which it has been tested, and it is often a feature of social science (as distinct from natural science) that the context of a theory can change so much over time that this alone means that new tests are always required. The distinction between inductive and deductive approaches divides statistics from mathematics as we have already noted in Chapter 3 above.

The second major contribution of philosophy can, in principle, come from moral philosophy or ethics. Since at times the geographer is concerned with different policies, questions can be asked about the 'values' of these alternatives and this obviously raises major questions. These range from issues of whether values have been built into certain theories surreptitiously or whether value-rules, which may be acceptable as general rules, can be applied to specific geographical problems. Examples of the latter are utilitarian principles ('the greatest good for the greatest possible number'), Pareto optimality ('changes are all right provided no one is made worse off') or Rawlsian optimality ('changes are all right provided the gainers are prepared to pay sufficient compensation to the losers'). More generally, if an equitable spatial distribution of some facility is sought, what is meant by equity?

This is not a philosophy text and so is not the place to argue out these questions in detail. There are many different philosophical starting points and positions. However, a general conclusion can be put – albeit as an assertion for the reader to investigate in his or her own philosophical reading – that there may well be an eclectic route forward: not to have to evaluate positivism versus phenomenology versus structuralism or whatever, but by critically understanding what each has to offer, to make the best use of *all* positions. We proceed, therefore, by stating a number of simple propositions (assertions) which transcend particular 'schools' but which reflect the positions reached in contemporary theorising. This provides the basis for a discussion of a number of important individual topics, and this will lead us to a position that is essentially *pragmatic* in Peirce's sense (see

e.g. Gallie, 1952) and which is largely captured in the *critical theory* of Jurgen Habermas (1974). We can then begin to evaluate the implications for urban and regional analysis and proceed to a discussion of some examples.

New roots from contemporary philosophy and social theory

We can begin with five linked assertions which help us to take the argument further:

(i) *Knowledge is a social product*. This asserts that epistemology, our theory about knowledge, cannot be divorced from social processes. It is not 'given' from outside society. The recognition of this 'truth' is the first of a number of steps we can now take which enables us to 'modernise' – to get into working order and to integrate – the ideas which come from the different kinds of philosophical schools discussed above.

(ii) *Language is made up of elements and a 'rule system' with a representation of meaning, all of which has evolved socially*. This is a corollary of proposition (i). Since knowledge is encoded in language, and knowledge evolves socially, then language also will evolve socially. Again, it is not 'given' externally. This proposition includes Saussure's idea of structural linguistics, but also makes it dynamic: it has evolved over time and continues to evolve.

(iii) *Meaning is established through communication between people*. This is the first step in an articulation of the social processes involved in propositions (i) and (ii). This will provide the link between a kind of objectivity and individual subjective experience. In the jargon, this kind of social interaction is known as inter-subjective communication.

(iv) *Truth is not something absolute, but a consensus*. This idea (once we go beyond logical and mathematical tautology) puts scientific knowledge on potentially the same basis as any other kind of knowledge. It takes us beyond positivism and a given observation language (cf. proposition (ii) above). Most science is 'true' because in the processes of analysis and experiment, it is relatively easy to reach agreement about what is the case and what is not. In other fields, this is not so; and this makes these fields both more difficult and potentially more interesting.

(v) *Much language and discourse has an ideological content*. A proposition can be described as ideological if it is based on values which are not universally held and which underpin a position unconsciously and implicitly. Many of the propositions of neoclassical economics, and hence much theory in urban and regional analysis (both classical and contemporary), fall into this category. We need to be continually aware of this and to learn how to strip propositions of their ideological content. There are a number of methods which purport to achieve this; for example, the notion of 'deconstruction', which is sometimes taken as the basis of a post-structuralist school.

We now discuss successively, the various concepts and topics which are the keystones of these propositions.

Language and meaning

When we learn a language, the roots of our understanding of meaning lie first in ostensive definition – a mental recording of an association between a word and an observed phenomenon; and secondly in acquiring a 'knowledge' of the rules – both the grammar of the language itself and other rules which 'connect' various concepts. These 'other rules' are, of course, elements of 'theories'. This process of learning is, and can only be, a social one: through teachers and other people; though once the 'rules' are acquired, the process can be very rapid because the whole of literature of all kinds, and other media, become available – an extraordinary repository of past experience. Paraphrasing Mary Hesse (1980), this all means that data, and 'meaning' more generally, are not detachable from theory; theories are the way the 'facts' are seen; language is irreducibly metaphorical and inexact.

At any moment in time, we have a current 'understanding' of language and theories about the world; but this understanding is constantly shifting, being re-appraised in the light of experience and inter-subjective communication. Much discourse inevitably consists of trying to establish mutual agreement about the terms to be used. And it is helpful in this to note Quine's principle of charity, to 'try to translate sentences of any alien language, particularly the observational ones, in such a way as to make as many as possible come out true in our language'. Note that an 'alien language' in this context is simply another person's. (Quine (1960) was citing N. L. Wilson (1959) in this context.)

To complicate matters, we need to remind ourselves continually that many of the 'theory-laden' sentences of any of our languages are likely to have an 'invisible' ideological content; and this is an issue to which we return shortly.

Truth

We have stated the proposition that 'truth' is what is agreed as a matter of social consensus. We can still retain the idea that we can seek after truth; but rather than comparing propositions to 'observations', we now have to relate them to the experience of ourselves and others, and to test the meanings in the languages in which we make the comparisons in the context of a wider social discourse (with continual interpretation and reinterpretation of terms). This is more messy than traditional positivist philosophy, but much more powerful. It can achieve everything that could previously be achieved: if it can easily be agreed what an 'observation' is, and how this compares to various propositions, then so be it – and that is how most 'science' is practised. But it also gives us the basis of describing what is happening in the social sciences and the arts, including human geography, when distinguished practitioners disagree. It provides an indication of how discourse can continue until either agreement is reached; or there is clear disagreement on interpretation (probably clouded by different views on ideology) or involving different value sets. Once we move beyond natural science, we should not be surprised to see this happening.

This kind of picture also allows us continually to work towards, and to focus on, truth and interpretation; and to argue that these epistemological questions are more interesting than the ontological issues on which these discussions sometimes founder.

Habermas and the three cognitive interests

Many of the points made so far can be neatly drawn together and extended by introducing Habermas's characterisation of an individual's subjective knowledge in relation to three kinds of cognitive 'interests': technical, practical and emancipatory. He sees technical interests as related to 'work' and the empirical–analytical sciences, practical interests to social interaction, and emancipatory interests to power. In relation to the earlier discussion, the 'technical' corresponds to the end of the knowledge spectrum where meaning and truth are relatively easily handled; the 'practical' to inter-subjective communication; and 'emancipatory' interests to ideology and power. This has the advantage that Habermas, unlike many social scientists (and including many geographers), is not out to attack science *per se*, but can locate it in its proper place. The substantial addition comes through linking emancipation with ideology. This aspect is admirably summarised by Habermas, as quoted by Bernstein (1976):

> The systematic sciences of social action, that is economics, sociology and political science, have the goal, as do the empirical–analytic sciences, of producing nomological knowledge. A critical social science, however, will not remain satisfied with this. It is concerned with going beyond this goal to determine when theoretical statements grasp invariant regularities of social action as such and when they express ideologically frozen relations of dependence that can in principle be transformed.

It would be nice to see human geography added to the list of relevant disciplines!

Theories

We have so far focused on individuals and social discourse and language, meaning and truth. We have remarked from time to time how propositions are 'theory-laden'. We must now turn our attention to the task of investigating what constitutes theories and seek to extend the argument about truth and understanding that was discussed in Chapter 3.

A theory is a complex of related propositions each more or less well-tested. There may be many different ways of formulating a theory but, given charitable translation, these can be thought of as the same theory. The objective of a theory is to 'explain' a set of phenomena; possibly to make predictions. It is useful to think of theories which represent different 'levels of explanation' in relation to such a set. Then, what we ordinarily think of as 'description' should be seen as a low-level theory. A research aim is continually to seek deeper levels of explanation.

Within a discipline or subdiscipline, it is often useful to think of theories as forming a network of related ideas, and authors like Quine and Hesse give considerable emphasis to the *coherence* then generated and the ways in which new ideas have to fit into this. (Though occasionally perhaps a whole system could be overthrown in a Kuhn (1962) type way!) Hesse also notes what the presence of value-determined coherence conditions will be once we move beyond the natural sciences, with connections to the notion of ideology again. She also notes that, typically, theories are underdetermined: there will be more than one theory compatible with available tests and it is more profitable to think of theories as being constrained by tests rather than being determined by them.

Values, ethics and politics

In the same way that 'scientific' propositions can be related to theory-laden descriptions of 'observations', so perhaps ethical terms can be related to theory-laden descriptions of pain and pleasure in individual experience. The 'theory-ladenness' in the second case will partly relate to social 'conditioning' and 'expectations', and partly to the ideological content of the descriptions when we begin to use terms like 'good', 'bad', 'freedom' and 'equality'. Many of the descriptions will reflect mutual power relationships of the social interactions which have generated them. We have seen that the 'observation language' cannot be taken as externally given; to an even greater extent, this is the case for ethical systems.

However, we can argue that there is nothing intrinsically different about value terms: they can be related to individual subjective experience and inter-subjective social communication in the same way as other elements of the language. We must, however, give very great emphasis to exploring the concepts of power and ideology whenever value terms are involved; and for the social sciences it can be argued that this is most of the time.

Concluding comments

This section provides an important backcloth to the approach adopted in this book. It provides a robust view of 'science' within a broader social context and thus enables modelling to be seen for what it is: as very successful when it is operating in areas where there will be consensus (perhaps demography being the least controversial example) but models of transport flows could be included here; and as a tool for developing and operationalising theory in areas which may be more controversial. What it should *not* be is identified with particular philosophical position (such as 'positivism').

KEY IDEA 4.2

Consensus in social science is not as easily reached as in, say, the physical sciences; modelling methods can be useful within what should be understood as a value-laden context.

STATISTICS, MATHEMATICS AND COMPUTER SCIENCE

The other obvious methodological disciplines are *statistics, mathematics and computer science* which we began to discuss in Chapter 3. Statistics is concerned with the analysis and interpretation of data and with the idea of inference, within confidence limits, of general laws from data in an inductive fashion. It also makes a direct contribution to deductive mathematical modelling through the estimation of 'best fit' parameters in those models and a calculation of 'goodness of fit'. It can play an important role in classifying the entities of systems, particularly through offering

rules as to whether certain observed differences between groups are 'significant' or not.

Mathematics will be approached in this book as a provider of tools for handling complexity. We have already noted that geographical systems and associated theoretical questions about those systems are very complex and it often turns out that 'understanding' or 'explanation' can only be achieved by resorting to the tools of mathematics. We discussed the concepts and methods which are needed for geographical theory in a preliminary way in Chapter 3. It is appropriate here, however, to make a number of general points about the use of mathematics in urban and regional analysis.

It can be argued that urban and regional analysis, through its constituent disciplines, changed its character substantially in a process of mathematicisation which began in the 1950s and reached a peak in the 1960s and early 1970s. To some extent this process is still continuing, though now more slowly. In geography, for example, the original change was often described as the 'quantitative revolution'. It can be argued, however, that the change is more of a revolution in the approach to theory: there is a greater will to grapple with all the complexities of theory, and these complexities happen to involve the use of mathematics. These developments, in a supradisciplinary fashion, have more recently been gathered together under the banner of complexity theory, which will be reviewed later in this chapter. An unfortunate corollary of these developments has been that much of the literature has become unintelligible to many students and practitioners within many of the disciplines associated with urban and regional analysis. This shift mirrors the shift from a comparative static mode of analysis to a dynamical one. What we will see is that a full dynamic analysis was not possible until the relevant mathematical tools became available. Indeed, it seemed (implicitly!) during the 1960s and early 1970s that such developments were unlikely. There was a focus on 'exogenous' variables which had to be fixed outside the model. These were the ones which could not be modelled dynamically. The situation changed with the publication of Thom's (1975) book on catastrophe theory. At first, the new insights were qualitative, but later a full mathematical approach was developed. A detailed account is given in Chapter 6.

It does turn out to be the case that many of the differences in approach to the theory of cities and regions can be associated not with differences of assumption or method, but with differences in the mathematical representation used for the geographical system being studied. This is particularly true for the issue of spatial representation first raised in Chapter 3. It turns out that more powerful mathematical techniques are available when discrete zone systems are used.

There are two other disciplines or subdisciplines that are associated with mathematics which have proved of great importance in different ways. One is computer studies, since the development of computer models is now an essential feature of geographical theory. The second is the branch of applied mathematics which is concerned with *modelling*. Increasingly, this is becoming the key methodology in relation to a wide range of systems and is, of course, the key theme of this book.

The development of modelling has been closely linked to computing power, and hence to the third discipline considered here – *computer science*. We should simply note here that while the ideas of computer science are connected to those of mathematics, particularly in the modelling area, they also add something distinctive. We have already noted earlier that computers can be used to simulate solutions to equation systems which would otherwise be mathematically intractable. This is accomplished through the application of computer algorithms; and more generally, we should consider what *algorithmic thinking* can offer (cf. Davies and Hersch, 1981; Harel, 1987).

The three disciplines of statistics, mathematics and computer science are, of course, conceptually linked, as we have implied above. These links become explicit in fields like neural computing (cf. Aleksander and Morton, 1990; Deco and Obradovic, 1996; Golden, 1996). For example, principal components analysis, entropy-maximising models, the calibration of mathematical models and the construction of new models can all be linked through neural computing. We will tease out some of the productive consequences of these links in Chapters 6 and 8.

SYSTEMS ANALYSIS

The final methodological discipline to be raised at the outset is one that is rarely taught as such, certainly not as a unified school or degree course, but which has had a profound influence on many disciplines. The argument of this book has been implicitly built on it through the concept of *systems*. Is there a discipline such as 'systems analysis' (or 'systems theory')? Many would argue so. The approach has a long history (cf. Buckley , 1968; von Bertalanffy, 1968; and more recently, in geography, in Wilson, 1981a). The subject also had a serious push in the late 1960s with the systems dynamics school of Forrester (1968, 1969). Recent examples of this style of work are provided by Hannon and Ruth (1997) and Ruth and Hannon (1997).

Systems theory has never been as successful as its advocates would have liked, probably because it is difficult to strike the balance between what can be achieved in general (*general* systems theory) and what needs the specific knowledge of a particular substantive discipline. The claims of the systems theorists have always been too great in practice. However, the reason why systems' concepts are given such prominence here is that in urban and regional analysis (and this is probably true of many other fields), there are important lessons to be learned which have been neglected in the past. The compromise which has to be achieved is to learn to take the maximum in terms of ideas from systems theory and any related enabling discipline (or indeed, substantive discipline where useful analogies can be invoked). The present author has pursued this issue for geography in the context of, as in physics, 'dreams of a final theory' (Weinberg, 1994; Wilson, 1995). In the future, systems theory is likely to be absorbed under the umbrella of complexity theory – or vice versa!

Systems analysis puts a very useful emphasis on interdependence of the components of a system and provides some tools for handling complexity. This, at the lowest level, is useful in a pragmatic sense. Secondly, it draws attention to the possible existence of 'systemic' effects: these arise when the behaviour of a system is very different from that which could be envisaged from a straightforward examination of its components. Systems of interest in urban and regional analysis have degrees of complexity and interdependence sufficiently great to make this kind of phenomenon likely. Thirdly, the practitioners of the discipline do argue (at least implicitly) the possibility of generality: that methods applicable to one kind of system may be in principle applicable to others of, in some sense, a 'similar' type. If this kind of programme of systems analysis had been fully worked through, then it would be possible to take any geographical system, identify its 'type' in some way, and then apply the methods for theory-building that are known from 'general systems theory' to be relevant. Of course, the intellectual programme involved has not been achieved, and perhaps never will be. But it is the case, as we will see later in this chapter, that there are commonalities between systems in different disciplines (which include geography) and that it is sometimes important for geographers to look to these disciplines for ideas and methods for theory development. What we learn about this kind of process from systems analysis it that it does not necessarily mean that we are then working by analogy: it may simply be the case that high level methods are being applied in a range of disciplines each of which have systems of interest which have enough in common to make this possible. This sort of notion is certainly a fruitful source of ideas for theoretical development in urban and regional analysis.

Finally, we note that 'systems theory' has different connotations in different disciplines. It is often seen by contemporary social theorists as 'functionalist' in its approach; suitably interpreted and developed, it can offer much to structuralist thinking.

THE NATURAL SCIENCES

In our discussion of the role of concepts from economics and the social sciences (and some related disciplines) it could be argued that the relationship to urban and regional analysis arose from a commonality of subject matter. This is not true of the natural sciences. In this case, any relationship is through systems analysis and the antecedents of complexity theory. In Chapter 6, we will provide a key example where a modelling technique arose because a similarity was discovered between its application in physics (which originated in the nineteenth century) and transport modelling. In this case, analogy provides a source of ideas. The applications of these ideas then have to stand up within the fields in which they are applied. However, in a number of contexts, it is possible to see the concepts involved as supradisciplinary and there is the exciting prospect of constructing *general* model-building tool-kits. The temptations of general systems theory again! In this sense, the natural sciences can function as an extension of the other enabling disciplines.

> ### KEY IDEA 4.5
>
> Concepts are under-used in the sciences. Those discovered to solve a problem in one discipline sometimes have a use in another. It is valuable to develop ways of thinking to characterise problems in general terms and then concepts can often have a supradisciplinary role.

COMPLEXITY THEORY

Introduction

Complexity theory combines much of the thinking represented in the preceding subsections in the context of the application of methods to the study of complex systems. Coveney and Highfield (1995, p. 150) write: 'Complexity arises at many

> ### KEY IDEA 4.4
>
> The concept of a system forces us to take interdependence seriously. Systems analysis is always a valuable starting point.

Table 4.1 Examples of scales

Physical
  subatomic, atomic, molecular, chemical, materials
Biological
  molecular, cell, organ, organism, ecosystem
Psycho-social
  individual, group
Economic
  individual, firm, industry, economy, trade
Urban and regional/geographical
  individual, neighbourhood, city, region, country,
  region

levels in nature, fashioning patterns and endless tiers of design.' We should add that this is just as applicable to the social sciences! And we should particularly note their reference to 'many levels', which connects to notions of scale and hierarchy introduced earlier. Examples of scale are shown in Table 4.1.

Table 4.1 shows how urban and regional systems fit into the broader scheme. From this starting point, we can explore the range of methods to be deployed in complexity theory, noting the possibilities of working by analogy if appropriate (as in the argument for the natural sciences above).

The measurement of complexity

A number of authors offer different measurements of complexity. Possibly the first was in a famous paper by Shannon and Weaver (1949) in the context of telecommunications traffic. They related the measure to entropy, which has an intriguing connection to the deployment of that concept in model-building to which we will return in Chapter 6.

For present purposes, a definition first introduced by Ashby (1956) will suffice – and this has the advantage of connecting model-based analysis to a critical idea in planning. He measured the complexity of a system by the number of possible system states. A coin-tossing system, for example, would have only two possible states: heads or tails. It is clear intuitively that this number rises – and to astronomical proportions – with

system complexity. We can simply bear this in mind as an informal notion. The connection to planning is a more technical one. It arises from Ashby's concern with control theory in engineering, and again the theorem has an intuitive plausibility: that the controlling system has to be at least as complex as the system that it is trying to control if it is to have any chance of succeeding. It is left, at this stage, as an exercise for the reader to think through what this means for the apparatus of city planning, for example, in relation to a city!

Casti (1995) adopts a different approach which is analogous, in a sense, to the Weaver approach introduced in Chapter 1. He notes characteristics of simple systems, implicitly defining complexity as the opposite. Simple systems are predictable, have few interactions and feedback loops, may be based on centralised decision-making and are decomposable. Complex systems are more likely to generate surprising behaviour. At the end of his book, he reaches a conclusion which is helpful in broad terms in relation to the approach of this book: '... the creation of a science of complex systems is really a subtask of the more general, and much more ambitious, program of creating a theory of models. Complexity – as a science – is merely one of the many rungs on this endless ladder.' This puts into context the aims of this book: to show that within urban and regional analysis, modelling and complexity theory have much to offer – but as a part of a larger whole.

The evolution of order

One of the main areas of interest in complexity theory is how order (usually at high levels of scale or in the hierarchy of perspectives) can *emerge* in such complex systems. A relatively ordered urban structure can arise, for example, out of the rich complexity of the interactions of all the individual agents in the city. Scientific examples include those which arise in chemistry, the human brain and astronomy. Prigogine (1980) was awarded the Nobel prize in chemistry for showing how order 'far from equilibrium' can arise in complex chemical systems (see also Prigogine and Stengers, 1984). (This work was earlier developed in an urban modelling context – for example by Allen and Sanglier, 1972.) Kauffman (1993, 1995) is concerned with the same questions for biological systems. (And of course, there is a long history of this kind of topic in developmental biology; cf., for example, Thompson, 1942; Varela, 1979.) Favre *et al.* (1988; p. 133 in the 1995 translation) are concerned with how structures can be modified through divergence from equilibrium.

These ideas have now been developed more formally within the context of economics through the work of the Sante Fe Institute (Anderson *et al.*, 1988; Arthur, 1994a, b) and Krugman (1993, 1995, 1996). Some of Krugman's work is directly on cities and regions (especially Krugman, 1996), as is some earlier work of Arthur (1988). These ideas have been developed in the context of the *emergence* of cities by Page (1998), with some connections to the cellular automata notions used by Holland (1995). A particular theme developed by Arthur is that of *increasing returns to scale* and that it is this property which leads to multiple equilibrium solutions in models of urban and regional structure. (We pick up the mathematical significance of this in the next subsection.) Arthur makes the important point that the *increasing returns*' assumption is avoided by most economists – hence the tendency in most economic models to generate single equilibrium solutions which do not accord with reality. It will become apparent in Chapter 6 that the increasing returns' assumption is anticipated in the locational models presented in this book. A recent review which covers a range of disciplines is offered by Ball (1999). We will attempt to integrate all these ideas, at least in the context of a research agenda, in Chapter 8.

It will turn out to be the case that even when we can explain how order can arise, there is a very great variety of possible structures in these various systems. Barrow (1991) has very interestingly raised the issue of how this can arise when the 'laws' governing the system may be relatively simple. He argues that it is in large part because of the variety of possible initial conditions – in effect that this is the source of another kind of complexity. His concerns are mainly the natural sciences, but we will be able to illustrate his ideas in the context of urban and regional analysis. There is a direct connection with Arthur's ideas here: because there are, typically, many possible equilibria, the actual state of the system will be determined by the historical path; this is another way saying 'is determined by the initial conditions', the 'path' being a time series of initial conditions.

Nonlinear mathematics

These ideas about complex systems can, of course, be represented mathematically. The essence, as we have already observed in an earlier context, is the nonlinearities within complex systems. When we solve a set of linear equations, the solution, if there is one, is unique. If the equations are nonlinear, it turns out that there are typically very large numbers of solutions. This connects formally to Barrow's point: the actual solution will be determined by the initial conditions. We would usually begin with a search for equilibrium solutions. But an additional feature which emerges is that there are values of the equations' parameters which are such that there is no achievable equilibrium: systems either oscillate or have a *chaotic* motion. And, of course, this has led to the notion of chaos theory rather than complexity theory. However, this is seriously misleading: for many systems, and probably most urban and regional systems, it is the multiplicity of equilibrium states that is most interesting. There is one other important feature of nonlinear mathematics to which attention should be drawn at this stage: that parameters in the equations turn out to have *critical values* at which there can be a discrete jump from one kind of equilibrium state to another – and this will be a major feature of urban and regional systems.

In the face of this kind of complexity, we can look to mathematics for insight. To solve the equations, again as noted earlier, we have to resort to computer simulation (for illustrations, see Levy, 1992; Kauffman, 1995; Casti, 1997).

Analogues and metaphors

Much of the core of what follows will be based on analogues and metaphors, though complexity theorists, in their neglect of the social sciences, have not had most of these ideas in their tool-kit. This will emerge as the argument develops in Chapter 6. What is important here is the extent to which we can look to complexity theory for new ideas which can be applied in urban and regional analysis.

Table 4.2 Possible analogues to use in model building

Genetic algorithms	Leung (1997)
Neural network computing	Johnson and Picton (1995), Deco and Obradovic (1996)
Ants	Coveney and Highfield (1995)
	Gordon (1995)
Simulated annealing	Davies (1987), Kauffman (1995)
Physics	Bak (1997), Jensen (1998)

Ideas to be explored include those shown in Table 4.2.

The possibility of a general model for complex systems?

One way of summarising the argument is to turn to the work of Holland – admirably summarised in his 1995 book. He tackles the issue of building general models for what he calls complex *adaptive* systems (cas). His main underpinning examples are biological or ecological, but he also uses 'cities' as an example, and this provides some insights for this book. He considers systems to be made up of *agents* which can *aggregate* in various ways to form meta-agents (or, for us, subsystems). He is particularly concerned with what we have identified as the scale issue, and in seeking to build an initial general model, he looks for a *two-tier* one. He also pays particular attention to the *geography* of his systems through the concept of *sites*. He then identifies four properties and three mechanisms which have to be incorporated into a general model to provide it with the richness to represent complexity and particularly adaptation. The properties are as follows:

- aggregation
- nonlinearity
- flows
- diversity

and the mechanisms are

- tagging
- internal models (for anticipation and prediction)
- building blocks

His agents 'own' and transform *resources*.

It is easy to see how to represent urban and regional models in this framework in terms of properties, and the mechanisms will be an important addition to our research agenda in Chapter 8. In Chapters 2 and 3 we have in effect been concerned with what Holland calls aggregation: forming meta-agents (such as retailers, to use an example that will be archetypal below) from prime elements; and cities certainly reflect diversity. We have already discussed nonlinearities. It is interesting and important in the context of this book that Holland gives such emphasis to the concept of *flows* – what we have termed interaction, and which we have argued is at the root of the urban model-building programme.

Holland's *mechanisms* are more concerned with evolution than with development. There is no difficulty with the concept of tagging – because we have been at pains to characterise elements and agents in cities and regions in rich ways which are consistent with Holland's concept. The 'internal models' are at the core of any learning mechanism, and this has always been a difficult area for urban modellers (see Wilson, 1975, for an early attempt at progress). The 'building blocks' are concerned with coherence of image when agents can be assembled in so many different ways.

These arguments are further developed in the context of emergence in his latest book, entitled *Emergence* (Holland, 1998). Compared to the earlier book, he gives particular attention to examples of mechanisms, and how emergence can be a product of quite simple mechanisms. Some of these may have direct applications in urban and regional analysis, as we will see in Chapter 8.

What Holland's argument does show is that the approach to urban and regional modelling represented here is at the heart of complexity theory as represented in urban and regional analysis. We will return to the deeper questions which are raised here in Chapter 8 when the foundations of what is currently known have been more fully articulated.

Concluding comments

Complexity theory is clearly going to continue to develop as a rich field. For urban and regional analysis, it offers insight from the kinds of application which have been accomplished in other areas; and some specific techniques. As a distinctive and important field, urban and regional analysis also has something to offer in return: first, a range of examples which may well offer analogues for other fields such as ecology; and secondly, as noted earlier, a demonstration that the ideas of complexity theory can be made to work for real systems.

The professional disciplines

ENGINEERING AND TRANSPORT STUDIES

The development of the economy, which is an important backcloth for much geographical theory, is in part driven by technological development, and so some knowledge of developments in engineering is important. However, there are also more specific connections, particularly through civil engineering, since this is the discipline within which much of the field of *transport studies* has been developed. Since, as we have argued earlier, much of *geographical* theory is concerned with spatial interaction, there are important contributions to be obtained from that source. Some transport engineering work is of a technical nature, for example, concerned with the construction of transport facilities – their costs and operating characteristics. But much of the development of transport modelling has taken place within the civil engineering and related professions and, partly for this reason, has not been fully incorporated into urban and regional analysis or geographical theory. We will attempt to ensure, in the account which follows, that this integration is achieved.

ARCHITECTURE, PLANNING AND DESIGN

Cites and regions have a physical infrastructure much of which is designed by architects, and this has led some architects to have an interest in urban form. March (see his chapters in Martin and March, 1972) was a particular exponent who linked the skills of architectural design with urban modelling. A leading member of his research group was Marcial Echenique, who has led one of the major groups to continue the development and application of integrated models, as we will see in Chapter 8 below (see Echenique *et al.*, 1990). With these kinds of exceptions, however, architects who have moved into urban design have generally done so under the heading of town (or city) planning.

Town planning is obviously a multidisciplinary exercise though it has forged itself into a professional discipline, mainly since the 1940s. Everything that has been argued about urban and regional analysis in this respect can be applied to town planning. Indeed, in one sense, urban and regional analysis is a subset of town planning: town planners at least need the analytical skills to underpin their work.

As a discipline (albeit a composite one), town planning had its origins in the work of pioneers like Ebenezer Howard who were driven by social idealism as much as technical skill. Most practitioners have come into the discipline from other disciplines, such as architecture, geography, civil engineering or economics and this mix is reflected in what has been achieved. However, the discipline has probably been dominated by its statutory responsibilities. These vary in different countries, but are essentially concerned with the control of land use (in the UK context, the granting of planning permission for development) and in many instances, of course, this is a valuable commodity. It should be possible to make these kinds of decisions on the basis of the best possible analytical capability but it has proved difficult to marry the two kinds of skills. And of course, town planning has suffered in the way urban and regional analysis itself has suffered since the late 1970s: any notion of 'planning' has been out of fashion and solutions, where possible, should be left to the market. This has led town planners to find other modes of employment – notably in the area of economic development. At this point they can link to the smaller group of planners who have typically worked at the broader *regional* scale.

There is, however, a very important link between the mainly qualitative accounts of urban and regional development which underpin the planners' understanding of cities and regions and model-based urban and regional analysts. The best of the analysts in such fields – and Peter Hall (1982, 1988, 1998) is an outstanding example – provide the benchmark against which the understanding that can be achieved from models can be tested. This thesis is explored in Chapter 8.

A final broader comment is appropriate. In Chapter 3, we drew distinctions between the activities of policy, design and analysis in different kinds of disciplines. Architecture and town planning can be seen to be disciplines that have a particular focus on *design*, especially perhaps the former. This is valuable in its own right and it exposes the issues of how analysis connects to design and indeed how to do an unpack 'design' as a concept. We will look at the first of these issues by exploring how models can be used in this way; and for the second, the recommended text is that by Alexander (1964), already mentioned in the context of Chapter 3.

SOCIAL ADMINISTRATION

At a more down-to-earth level, we find the practitioners of 'social (or public) policy and administration'. They again have a common interest in the basic theory, shared with the other social science disciplines, but are also specifically concerned with articulating and understanding the vast array of government policies at all levels and their impacts on society. Their work is a rich source of material for geographers whose (yet more specific) interest would be in the spatial impacts of that broad range of policies. On the policy–design–analysis spectrum reintroduced in the previous section, they are very much concerned with the policy end of the spectrum, though not operating fully comprehensively. They are particularly important in relation to subsystems such as the housing system.

Integrated disciplines: urban and regional studies

There was a period in the 1960s when it seemed that what might be the core disciplines with an interest in cities and regions were not fully taking into account the new developments in urban and regional analysis; nor indeed were they offering much in design and policy terms to help confront what were being recognised as a set of very serious problems such as the decline of inner cities. A number of universities established Centres for Urban Studies and journals such as *Urban Studies*, *Regional Studies* and *Environment and Planning* were established, and there were corresponding professional or academic societies, the most influential of which for urban and regional analysis has been the Regional Science Association, founded by Water Isard in the 1950s (which has generated its own array of journals). On the whole, the centres have not survived while the journals and the societies have, and this may say something about the nature of interdisciplinary studies: there will be cases where the new coalition does establish itself as a new discipline – and interestingly, this may have been achieved on the narrower front by *transport studies*; there will be cases where this does not happen, but where the coalitions are valuable and the societies and the journals provide the media for these to work; and there will be cases where nothing survives! Urban and regional studies has achieved the intermediate status.

Concluding comments

Disciplines are interesting social constructs. We have referred to a great variety of them in this chapter, each of which either has cities and regions as one set of objects of study (usually among may others) or has a methodological contribution to make to urban and regional analysis. Two points can be emphasised. First, the multiplicity of disciplines is relatively recent historically – a process which started in the late nineteenth century but which only accelerated in the middle of the twentieth. Secondly, while most disciplines can be defined in reaction to their subject matter (cf. Wilson, 1992), they are social coalitions and the extent to which new ones can develop or survive depends on their relative strength in this milieu.

Cities and regions, with all their subsystems and complexities, are connected to an unusually wide range of disciplines, and this alone may be one reason why it has not been possible to focus attention within a new coalition. However, there may also be advantages in this. Provided there are enough people who are prepared to take an integrated view, and are also prepared to search systematically all other disciplines for any concepts or developments that can help urban and regional analysis, then there is strength in this diversity. There is a wider range of resources to call on.

The professional and academic societies, and the journals, therefore play a critical role in holding this looser non-disciplinary coalition together. We can say with confidence that all the disciplines reviewed briefly have something to offer urban and regional analysis; but we can say with equal confidence that the reverse is also true!

5

Classical models

Introduction

Where should complex spatial systems' theorists or urban and regional analysts look for classical models? Since cities and regions are central to human geography, since geographers have attempted to absorb the appropriate influences of others disciplines, and since an historical approach is not only interesting in its own right but provides a context, we can take geography as a suitable starting point – though the authors would not usually call themselves geographers. Which authors might we then deem to be the 'classics' of geographical theory and modelling – and hence of urban and regional analysis? We have to include at least von Thunen and agricultural land use; Weber and industrial location; Christaller and Losch and central place theory; Burgess, Hoyt and Harris and Ullman in relation to residential land use; and the early spatial interaction modellers such as Ravenstein, Reilly and Zipf.

The choice can be made simply on pragmatic grounds: they are discussed in most geography textbooks; but we can quickly see that this range of examples can be justified in other ways. Between them, they offer an amazing diversity of approach and they cover most of the main types of substantive systems of interest. (We can associate 'services' with both central place theory and Reilly among the spatial interaction modellers.) They cover the location of both dispersed activities (agriculture, residential location) and point activities (industry, services, settlements, spatial interaction). There is relatively little on *change*; they are mostly more concerned with explaining

structure at a particular time, though limited exceptions are Ravenstein on migration and Burgess and Hoyt. They also show the influence of different disciplinary outlooks: von Thunen, Weber and Losch were essentially economists, Burgess a representative of early Chicago sociology which combined social science with ecology, and the 'gravity' modellers had roots in physics. Hoyt, Harris and Ullman and Christaller were all more pragmatic, geographers (mostly) in name and style.

All in all, between them, they cover the main geographical problems for which theory is in principle required, though this point is not usually emphasised as such when they are presented. It is important, therefore, that we understand what they achieved and that they each represent the source of streams of work, the courses of which we can follow. In particular, by scrutinising each in relation to the STM dimensions of Chapter 3, we can gain a better understanding of why the classical basis of geography came out as it did and the nature of the advances achieved in contemporary theory. In each of the following sections, therefore, we present the main ideas and results of each of these sample contributions, classify it in relation to the STM dimensions and offer a brief critical assessment and an indication of how it connects to future work. Because these contributions are in a sense the foundation of geographical theory, and because in most textbooks the presentation is very brief and uncritical, we offer a more detailed presentation and critique in Appendix 2. This permits some simplification in the presentation of the main results of these authors below.

von Thunen's rings

Von Thunen (1826) had a number of 'main' results, as Appendix 2 shows, but we concentrate here on one – the distribution of agricultural land use – crop types – around a single market centre. He calculated the potential profitability of each crop at each distance from the market – revenue from sales at the market minus production and transport costs – and then took the 'profit' per unit area of each crop as the maximum rent which a landowner could charge at that distance. Because of the workings of the land market, the most profitable crop would be sown at each location. This analysis leads to the familiar pattern of rings shown in Figure 5.1(b), with the most profitable crops nearest to the market centre. The core of the analysis can be stated in relation to the figure. Let Y be the yield per unit area at distance d; let p be the price per unit weight, v the unit production costs and c the transport cost per unit weight. Then the profit is

$$\Pi = Y(p - v - cd) \tag{5.1}$$

This can be worked out for various crops at various distances. The essence of von Thunen's idea is that the crop would be planted which gave the highest profit at any particular distance from the centre. When $d = 0$, this will clearly be the crop for which $Y(p - v)$ is a maximum. For convenience of notation, call this crop 1. There will come a distance, say d_1, at which another crop, which we call 2, will become more profitable; that is:

$$Y_1(p_1 - v_1 - c_1 d_1) < Y_2(p_2 - v_2 - c_2 d) \tag{5.2}$$

for

$$d > d_1 \tag{5.3}$$

and this analysis can be continued for successive crops (which for convenience we continue to number in order of decreasing profitability with distance) – and this is the essence of Figure 5.1a which leads to the rings in 5.1b. Its construction is explained in detail in Appendix 2.

We now consider each of the STM dimensions and their main subheadings in turn.

SYSTEM ARTICULATION

(i) The entities are agricultural products in relation to a single market. Implicitly, there are producers and consumers.

(ii) The level of resolution is relatively coarse: there are few crop types and there is the 'point' market, assumed to be a town. We should also note that von Thunen also worked out his theory for finer scales (see Appendix 2).

(iii) A continuous space representation is used: the objective of the theory is to demarcate boundaries between different land uses.

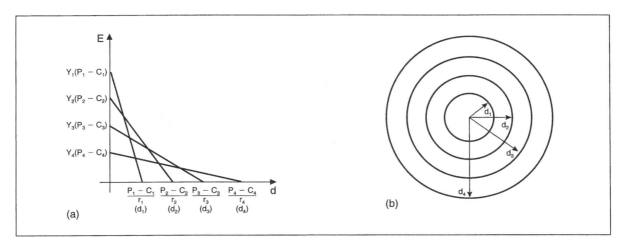

Figure 5.1 von Thunen's rings

THEORY

(i) The approach is essentially partial because of the assumption of a single market; and because individual farms are not identified. There is no explicit representation of competition, therefore, either between markets for goods or between farmers for markets.

(ii) The main theoretical basis is profit maximisation but with a brilliantly simple idea of how the land market works. Since the 'profit' which can be extracted as rent is seen as the modern concept of 'bid rent', then von Thunen laid the basis for much contemporary location theory. The surprise is perhaps that this idea was not developed further sooner than it was.

METHODS

The methods used were arithmetical and graphical. Essentially, what can be seen as a simple algebraic model (as presented in Appendix 2) was 'solved' for specific instances. Now, as we will see later, the ideas of mathematical programming can be brought to bear on this type of problem.

Von Thunen's work, therefore, has some brilliant elements and offers some important insights. It is essentially limited by the assumption of the single market. It is possible to begin to see how the land use pattern changes if there are, say, two markets, and if other assumptions (say about uniform fertility or transport costs) are relaxed, and von Thunen himself attempted this. But the formulation and methods do not allow us to proceed immediately to the general case of any number of markets in a regional or national settlement system. It will be argued later that there are essentially two reasons for this: first, and crucially, the adoption of the continuous space representation; and secondly the lack at the time of some basic mathematical techniques. Because of the first of these deficiencies, and subsequent researchers adopting the same assumption, the weaknesses have taken a long time to repair. Notwithstanding the critique, we should applaud the key idea.

Weber's triangle

A reading of *A Theory of the Location of Industries* makes it clear that Weber (1909) has much more to offer than the famous 'triangle' associated with his name in texts; but these topics are pursued in Appendix 2 and here we restrict ourselves to the triangle, shown in Figure 5.2, to illustrate the argument. This particular problem is as follows: given two sources of material inputs for a factory (M1 and M2 in Figure 5.2), and a single market for its products (P in Figure 5.2), what is the optimal location of the firm (X in Figure 5.2)? Weber argued that the only variable costs were transport costs, and so these had to be minimised to find the optimum. If d_i in the figure represents distance, and w_i represents weight, and transport costs are assumed proportional to weight and distance, then X is the point for which

$$Z = w_1 d_1 + w_2 d_2 + w_3 d_3 \qquad (5.4)$$

is a minimum.

For different cases, essentially according to the nature of the materials inputs (the extent to which one or more weight-losing materials are involved in the process, and the proportion attributed to

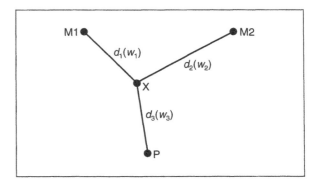

Figure 5.2 The Weber triangle

ubiquities), there are three kinds of optimum: location at one of the materials' sites; location at the market; or location at a point within the triangle. The details are worked out in Appendix 2.

What were Weber's decisions on the STM dimensions?

SYSTEM ARTICULATION

(i) The system of interest is the factory, the materials' sources and the market.
(ii) The level of resolution is obviously a fine one: individual organisations and exact locations are involved.
(iii) A continuous space representation is used.

THEORY

(i) The theoretical perspective is very partial: there is only one factory, and so there is no way of representing competition between factories for materials inputs; and there is only one market, so there is no competition for or between markets.
(ii) The theory is based on a profit-maximising hypothesis, which is translated through simplifying assumptions into a transport-cost minimising hypothesis.

METHOD

Weber was able to use a machine – the Varignon frame – for solving his problem. It is interesting that it is a relatively difficult mathematical problem and even today the triangle problem can only be solved heuristically and iteratively on a computer.

Once again, it will turn out that a shift to a discrete zone basis for the problem opens up Weber's model to powerful new mathematical techniques – those of mathematical programming – though it should be emphasised that these were discovered in the 1940s with the more general nonlinear results only being established in the 1950s and onwards, so these techniques were certainly not available to Weber! It should also be emphasised that Weber was well aware of the nature of his assumptions and the character of the more general problem. Much of his book can be interpreted as though it was written in the spirit of systems analysis.

However, still in the classical period – from the 1920s onwards – attempts were made to deal with some of the problems of building a theory of industrial location which Weber failed to tackle, and in particular the problem of competition when more than one firm is admitted to the picture. We therefore consider briefly the work of Hotelling (1929), Palander (1935) and Hoover (1937) to illustrate this point. Although Hotelling comes first chronologically, we consider his contribution last because its longer-term significance is greater.

Palander considered the problem of optimally locating two firms, producing goods for the same market. The prices of the goods at the point of production could differ, as could the transport rates. He was then able (as described in Appendix 2) to delineate the market area boundary between the two firms. The essential point to note for now is that his system is still essentially a simple one and, partly because of using a continuous-space representation and partly because of the economists' idea of cost-minimisation for the consumer, he focuses on market area and boundaries. Hoover extends the same kind of analysis to more than one firm and delineates market area boundaries through the device of drawing contours of equal delivered cost. (It is worth noting that in later years, Hoover (1967) worked with a computer model which anticipated many of the methods of dynamic modelling to be described in the next chapter.) Hotelling's problem (again described in more detail in Appendix 2) was the famous example of two ice-cream men on a linear beach, as graphically represented by Alonso (1960, 1964). Assuming even demand, the 'welfare optimum' is to have one seller at each quartile, thus minimising the consumer's average trip length. But then, one could move nearer to the centre and capture more of the market; the other, in defence, has to follow suit: the only *stable* equilibrium is where they are both at the centre. This parable says a lot about the nature of agglomeration and spatial competition. It was an active treatment of the interaction of two 'firms' focusing on the process which generates the final pattern – in contrast to Weber, Palander and Hoover. Once again, however, the focus is on market areas within a continuous space representation.

With all the examples considered, insights have been gained through the investigation of simple examples. No methods were offered which hold out any hope for dealing with real-world complexity. However, there is a key idea.

Christaller's hexagons

It is interesting that the work of nearly all the classical theorists can be characterised by a geometric symbol, and in the case of Christaller (1933) and central place theory, this is the hexagon. This arises, as explained in Appendix 2, by considering the intersection of *circular* market areas of settlements arranged on a regular lattice.

Central place theory is based on two main ideas and a number of assumptions. The first idea is that when consumers pay for a good they pay the production price plus transport costs, and the amount they purchase is determined by a demand curve of the usual shape. Thus purchasers at some distance from a centre buy a smaller quantity, and there is a distance, the ideal *range of a good*, beyond which no purchases are made. The second idea relates to the nature of the production function of any good. There is assumed to be a minimum threshold of market area to support the production of that good. From these ideas, we can begin to develop the notion of the *order* of a good: low range and low threshold correlate with low order; and vice versa.

The addition of some assumptions to these ideas then generates the well-known nested hierarchy of hexagons, as shown in Figure 5.3. The assumptions are concerned with a uniform density of an agricultural population as a uniform plain, with the transport system, and with the relationships between different orders of centres. The illustration here is a system based on the *marketing* principle; the other alternatives are generated in Appendix 2. The left-hand side of Figure 5.3 shows a number of point settlements which might be considered to supply low order services to the surrounding rural populations in each case. To the right, the next tier is shown: higher order centres now each serve six lower order centres. Further to the right, there is the next higher order set of centres; and so the arguments can continue to build up.

We now have just enough information to relate Christaller's ideas to the STM dimensions.

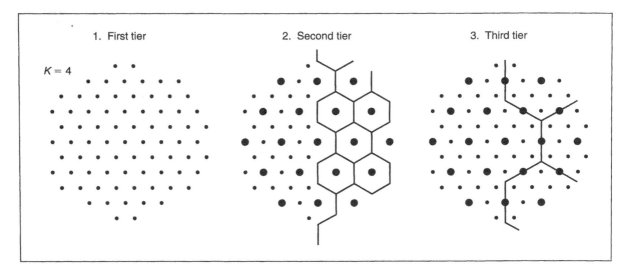

Figure 5.3 Examples of central place structures. (From Haggett, Cliff and Frey, 1977)

SYSTEM ARTICULATION

(i) The entities are point settlements with which are associated facilities for the production of goods; and a rural population.

(ii) The level of resolution is relatively coarse.

(iii) Space is again treated continuously, with a focus on the demarcation of market-area boundaries.

THEORY

(i) There is an obvious sense in which the approach is comprehensive, but rigidly so: it fails to offer any account of competition between settlements.

(ii) The theoretical base is relatively simple neoclassical economics, including, as noted earlier, the principle of non-overlapping market areas.

METHOD

The 'solution method' lies in the application of geometry and trigonometry. The uniform plain consequences can be relaxed by assuming the whole system to be on a rubber sheet which can be 'stretched' in various ways.

The great virtue of the Christaller system is that it systematically develops the clearly desirable concept of the *hierarchy* of settlements. But it is equally clearly too rigid. To construct the hierarchy of settlements, the spacing of settlements at *one* level in the hierarchy is available as a free parameter, and the fixing of this determines the spacing of centres at all other levels – up or down. The real world simply cannot be as rigid and neat as this. Thus we gain some insight, but from a theory that is ultimately unrealistic and simple.

As we describe in Appendix 2, Losch (1940) developed an alternative central place system that is less rigid and has a greater variety of market areas for different kinds of goods. However, that has its own oddities, particularly the principle of 'rotation' used to determine the final structure. Again, we have to recognise an interesting early attempt to grapple with a difficult problem; but for solutions, we have to look for new representations and methods, which will be presented in the next chapter.

More rings, with sectors and points: Burgess, Hoyt and Harris and Ullman

When we look at residential location patterns, it is not surprising that the first notable feature is a ring structure. This might well follow from a modification of von Thunen's argument for agriculture – and indeed, Alonso proceeded in such a way in the 1960s. However, the first investigator of rings, Burgess (1927), used a completely different theoretical basis: a combination of early sociology and the ecological theory of invasion and succession. Cities are growing, incomes grow, residents move out in search of better sites, and a higher income group comes to dominate an outer ring. New migrants fill gaps in the innermost ring. Hoyt (1939) showed how this theory had to be modified to incorporate sectoral differentiation. In the small-town beginnings of a city, the better-off would adopt the better environment. That sector would be then surrounded by other classes or uses. So expansion has to be outwards; and combined with sectoral differentiation in rings. Harris and Ullman (1945) took the argument a stage further by pointing out that expanding cities absorbed smaller towns and villages, and this alone imposed new centres on the overall structure. There were also Weber-like agglomeration economies in the supply of services. So we finish up with a picture of rings, with sectoral differentiation and additional centres superimposed.

We can now summarise these analyses in relation to the STM dimensions.

SYSTEM ARTICULATION

(i) The main entities are residential land use, though residential activity is to an extent linked to associated services.

(ii) The level of resolution is coarse, though to be fair, as we note in Appendix 2, sociologists such as Burgess simply used the ring structure as a broad framework for small-area studies.

(iii) All authors are essentially using a continuous space representation and their main concern is the demarcation of the boundaries which distinguish different kinds of areas.

THEORY

(i) The models are crudely comprehensive and, through the ecological theory, there is some notion of competition between groups.

(ii) The formal theoretical basis is ecological, but this does not have much substance on examination. The main theoretical foundations are 'low-level' and might be said to be descriptive–empirical; however, this should not be disparaged, and much insight can be gained in this way.

METHOD

No 'methods' are demanded beyond elementary cartography and geometry.

KEY IDEA 5.4

Since we can find qualitative hypotheses that generate a variety of different patterns, and since examples of all these patterns can be found, then we should be seeking a more general theory that can generate a composite pattern as a mix of the key features. See Chapter 7!

A mosaic of flows: elementary spatial interaction models

We move now for the first time from a geography of rigid geometries to the possibilities of a more complex pattern: a mosaic of flows, a matrix connecting any small area of a city or region to every other. For a city with 100 zones, there are $100 \times 100 = 10,000$ possible flows, each of which could be marked as an arrow, with thickness proportional to the flow, on a map. Sheer numbers usually inhibit this kind of portrayal.

The early interaction models were known as 'gravity models' because they were based on an analogy with Newtonian physics. The idea is attributed to Carey (1858) and was applied to migration by Ravenstein (1885), Lill (1891) and Young (1924) and to retailing by Reilly (1929, 1931). See Erlander and Stewart (1990) for a review. The models were formalised by Stewart (1942) and Zipf (1946) in the 1940s. Zipf, in particular, applied the concept to a wide range of flows. We note in Appendix 2, however, that almost all the authors applied the model in inter-city contexts: certainly to situations where the interaction could be considered to flow from point to point. The concept of discrete zones, and, as an approximation, flows being between zone centroids, was not made explicit. Even more remarkable was Reilly's use of the model: to demarcate the market areas of two points – essentially producing a retail version of Palander's industrial model. So although the whole logic of an interaction model implies the much needed recognition (which is empirically confirmed) that flows, and therefore market areas, overlap, Reilly was still attempting to use the model more in the style of other classical theorists.

We can conclude by summarising the approach in relation to the STM dimensions.

SYSTEM ARTICULATION

(i) The entities relate to whatever flow is of interest, and what produces or attracts those interactions.

(ii) The level of resolution is usually coarse (though as we will see in some of its later manifestations, this is not necessarily the case).

(iii) There is an implicit discrete zoning system, which much later was made explicit. In early applications, flows were between points and analyses were sometimes coupled with continuous space questions, as with Reilly's treatment of retail markets – boundaries of demarcation between areas.

THEORY

(i) The approach of these models is essentially partial; though the main development which produces contemporary models does handle competition and in that sense there is the basis of a comprehensive model.

(ii) The 'theory' is based on an analogy with Newtonian gravity:

$$I_{ij} = KP_iP_j/d_{ij}^2 \qquad (5.5)$$

for a flow between i and j, I_{ij}; 'mass' terms which were nearly always measured by populations, P_i and P_j; and an inverse distance relationship $1/d_{ij}^2$. The value of the exponent, 2, comes from the Newtonian analogy and there is no reason whatsoever for this to hold for economic– and social–geographical systems.

METHOD

The 'method' for model-building demands no more than simple algebra, though it is helpful to be able to 'picture' I_{ij} as the (i,j) element of a matrix:

$$\{I_{ij}\} = I_{11}\ I_{12}\ I_{13}\ \dots\ I_{1N}$$
$$I_{N1}\ I_{N2}\ I_{N3}\ \dots\ I_{NN} \qquad (5.6)$$

for an N-zone system. But more of this later.

There is a good argument for including interaction modelling as part of 'classical' geographical theory. Its chronology supports this; but it is also fascinating to see how the seeds of contemporary theory were present through virtually the whole of the classical period, though the products were largely used in the style of then-contemporary work rather than as the basis for a break into something new. And in this case, unlike that of, say, Weber, there was not the excuse that the relevant mathematical techniques were not available.

Concluding comments

We have now seen contributions based on clever ideas whose applications have been limited because of inadequate algebraic formulation or lack of appropriate mathematical techniques. The problems of algebraic formulation can be related to the use of a continuous space representation rather than a discrete zone system. Both von Thunen and Weber fall into this category. Christaller's central place theory is based on some economic principles, but the results turn on overly rigid geometrical assumptions. Burgess *et al.* are largely descriptive. The interaction modellers, as we have just been arguing, had the right idea, but failed to carry it through to its ultimate conclusion. We now argue that the interaction model, appropriately extended, can provide a basis for a much more comprehensive and unified theoretical basis for urban and regional analysis and that the different classical theories of this chapter can all be more effectively presented and developed as special cases of this formulation. This is the subject of the next chapter.

6

Interaction and location

Introduction

INTERACTION AND LOCATION: THE STM PERSPECTIVE

The central thesis of the book will be developed in this chapter. It is that by adopting an *interaction* modelling perspective in a discrete zone representation of cities and regions, it is possible also to build *location* models covering a wide range of examples. In terms of the STM framework, *system articulation* can embrace the main subsystems in turn: agriculture, industry, services, residential/housing, and transport. These elements can then be combined into a comprehensive model which, with appropriate resolution decisions, will also function as a model of settlement structure and hence serve as a replacement for central place theory. We present this full range in Chapter 7. Here, we focus on two examples: transport flows as an example of 'pure' interaction; and a model of the location of services, and in particular of retailing, which provides an archetype for location models. This enables us to introduce the key ideas which serve as a model-building tool-kit which the reader will be able to apply to his or her own problems – as well as to the examples in Chapter 7.

In STM terms, the levels of resolution used in the models will be determined in part by the discrete zone system. In sectoral terms, it should be sufficiently fine to identify persons and organisations by suitable categories; and in each case they should be located by zone within a discrete zone system. Typically, individuals will not be identified.

The *theoretical basis* for model-building is nearer to neoclassical economics than anything else, but typically uses economic concepts under the assumption of imperfect markets. From another point of view, the typical hypotheses adopted might be seen as simply 'descriptive' or 'pragmatic', but allowing, to some extent, economic interpretation. The essential point is to work with a sufficiently fine level of resolution to be able to formulate hypotheses that are adequate to represent the high levels of nonlinearities and interdependencies that are found in real systems. The approach is comprehensive in that it can at least handle competition between, say, a group of organisations in one zone and those of another; or, in suitable cases, between individual organisations across space.

The *methods* employed here are mathematical and link to the origins of complexity theory in the 1970s – not then of course recognised or acknowledged as such. The key ideas can be understood using the notation of relatively simple mathematics, mainly matrix algebra, though this looks complicated at first sight because of the presence of large numbers of subscripts and superscripts. A full account of the underlying theory demands a knowledge of calculus, mathematical programming and aspects of dynamical systems theory – what can now be seen to add up to the mathematics of complexity theory. We restrict ourselves here to an account which omits the mathematical detail and points in the direction of supplementary reading where appropriate. Readers who are unfamiliar with the kind of mathematical notation used here should consult Appendix 1 at this stage.

As the title of the chapter implies, the geographical foci of the systems defined are the *locations* of the various entities, *interactions* between them, and the more slowly-changing variables to represent *structure* – all of which enables us to track systems through time and to model *dynamics*. The interaction perspective is critical. The story of the development of this capability is an interesting one, and it is appropriate to begin with an historical review which situates these model-based developments within the context established by the classical theorists in the previous chapter. The dynamical models will be seen as archetypes of complexity theory, though were not recognised as such at the time of their development. They thus provide the basis for a theory of complex spatial systems.

THE ARCHETYPAL EXAMPLES

We can complete this introduction by describing the two archetypal systems to be used as key examples. Assume a zone system such as that shown in Figure A1.1 in Appendix 1. In the figure, we show a typical element of the matrix $\{T_{ij}\}$ as representing the journey to work: T_{ij} is the number of people who travel from residences in zone i to jobs in zone j. It will also be useful to define a variable to represent the total number of residents leaving zone i each day for work travel, as O_i; and the total number arriving in zone j, D_j. We use the variables O_i and D_j because we can take O and D respectively to stand for *origins* and *destinations*. The other key variable is some measure of travel impedance between i and j and we will usually define this to be a matrix $\{c_{ij}\}$ – the travel 'cost' between i and j. However, this should be thought of as a *generalised cost* made up of a number of elements. For example, we might have

$$c_{ij} = a_1 t_{ij} + a_2 r_i + a_3 s_j + m_{ij} \qquad (6.1)$$

where t_{ij} is the travel time between i and j, r_i represents a waiting time at i, s_j a terminal 'time' at j (e.g. walking time to final destination); and m_{ij} is the money cost of travel between i and j. In this representation, therefore, because m_{ij} is weighted as 1, the coefficients a_1, a_2 and a_3 represent the 'values' of different kinds of time and will be important parameters in the model. It is important to know

that the models can incorporate this level of detail, but henceforth, we will largely restrict ourselves to using the single variable, c_{ij}, as a travel cost.

It is possible to add sectoral detail. For example, we might want to model the number of persons of type n (say income or social class) travelling from i to j for work purposes by mode k. The matrix element T_{ij} would then become T_{ij}^{kn}. The trip totals at i and j would then become O_i^{kn} and D_j^{kn} respectively.

A similar portrait can be put together to represent a retail system. We take S_{ij} as the interaction, usually measured in money terms as retail sales, by residents of i in a shopping centre j. (In this case, the set of origin zones, $\{i\}$, and the set of shopping centres, $\{j\}$, are likely to be different.) Instead of total trip origins, O_i, we introduce a composite term $e_i P_i$, where e_i is the per capita expenditure by residents of zone i and P_i is the population. And in place of total trip destinations, D_j, we introduce a measure of 'attractiveness' (that which pulls shoppers to j) and call this W_j. As in the case of generalised travel cost c_{ij}, this will be made up of a number of factors, usually taken multiplicatively in this case:

$$W_j = (X_j^1)^{\alpha_1}(X_j^2)^{\alpha_2}(X_j^3)^{\alpha_3} \dots \qquad (6.2)$$

where α_1, α_2 and α_3 are parameters. As in the generalised cost case, we can bear this in mind, and it will become part of the richness of model-building potential; but we will normally use the variable W_j as attractiveness, possibly itself raised to a power, α – and hence as W_j^{α}. Typically, we will take size as a component of attractiveness (as one of the X_js) and sometimes, for ease of presentation, we will make an even simpler assumption that W_j is proportional to size, and that W_j^{α} is the attractiveness term.

An historical review: from the classical to the new

DEVELOPMENTS IN GEOGRAPHY

In the previous chapter, we reviewed, through the classical theorists, developments up to the 1940s. There was a new era of mathematical modelling as a basis for geographical theory from the late 1960s onwards. The period in between, the 1950s and 1960s, represented a time when urban and regional analysis was not formally developed; the nearest specialist discipline, human geography, became more systematic and there were the beginnings of a quantitative 'revolution' (the latter being first described as such by Burton in 1963). There was much conceptual and verbal-theoretical development, a greater recognition of the interdependence of elements through systems analysis (with thinking often usefully represented in complex diagrams), and, along with much data collection, the testing of hypotheses using statistical methods. Some of the concepts, like Haggett's (1965) *nodal region*, provided broader frameworks, and the elements of classical theory could be seen to contribute to them. Other theoretical developments were concerned with refinement of classical theory. There was much research on central place theory and city size distributions, for instance, and Berry (1967) applied the concepts of central place theory in an intra-urban context to retailing facilities. A major generalisation of the ideas of the socio-ecological school was established through the introduction of factor analysis (cf. Berry and Rees, 1969, Timms, 1971; Rees, 1979) and this deepened our knowledge of the social structures of cities. The evolution of the subject from its statistical beginnings to the kinds of mathematical models used in this book is well described by Robinson (1998).

ECONOMICS, REGIONAL SCIENCE AND PLANNING

Perhaps the most important influences came from outside geography. Some economists began to recognise the importance and interest of locational and regional problems and in the late 1950s, the Regional Science Association was formed as a multidisciplinary coalition, but firmly rooted in economics. City planning departments grew rapidly; some civil engineers concerned themselves with transport planning problems. The work of Isard (e.g. 1956, 1960; Isard *et al.*, 1969) shows that much was achieved. But many of the developments relied on the continuous-space representation and this had the effect of limiting much theory to patently unrealistic monocentric cities. Even within this paradigm, there were some brilliant pieces of work such as Alonso's (1964) book on residential location (and see Alonso, 1960) which applied von Thunen's ideas in a modern and urban context – though this was only really to achieve full fruition when translated to a discrete-space representation by Herbert and Stevens (1960); and the full significance of this in the wider modelling context was not realised for a number of years as we will see later (cf. Senior and Wilson, 1974). In Chapter 4, we noted that the work in the economic paradigm has been continued, partly in the context of complexity theory, by Arthur (1988, 1994a, b) and Krugman (1993, 1995, 1996). Arthur's work is not part of the traditional economic mainstream and is mainly based on the idea of increasing returns to scale. We will see later in this chapter that the dynamics of our core model can be interpreted in this framework.

The main post-war theoretical effort could be seen as firmly rooted in the classical tradition. Lurking within this were the seeds of a later revolution: the application of mathematical programming to location theory; and the models of the transport planners. This saw its expression in a variety of studies in the early 1960s which were rooted in discrete zone systems, comprehensive model-building attempts and some links to mathematical programming – in one case, Schlager (1965) combining modelling with optimisation and design. Important early papers include Harris (1962) and the special edition of the *Journal of the American Institute of Planners* in 1965. A summary of this early work is provided by Boyce *et al.* (1970). This early history forms a major part of the story that follows.

THE BASIS FOR MODELLING

We can recognise in the early history of modelling ambitious theorising, vast quantities of data, computer information systems and mapping and much statistical analysis, whilst still being largely and visibly rooted in classical theory. There may have been a quantitative revolution; there may have been an enormous expansion of geographical knowledge both inside and outside the discipline.

There was considerable theoretical development. But what was needed was some further integration that would bring about a theoretical as well as a quantitative revolution. Modelling provides the framework for this.

The basis for the modelling, with its roots in the 1960s, as has already been implied, is the use of a discrete-zone representation. The best ideas from continuous-zone theories can then be translated and integrated with the discrete-zone models (which had not been recognised early on, as being particularly significant for geography). This should be borne in mind, therefore, in the rest of this history and in the central argument of this chapter.

In the rest of this section, we focus mainly on the development of discrete-zone interaction and location models. One further discussion of preliminaries is necessary, however: this is partly about scale, and partly to avoid completely neglecting an important set of models. It has already been implicit in the book so far that there are perhaps two fundamentally different spatial levels of resolution (though, in the end, these should obviously be considered as points on a spectrum). These might be called the inter-regional (or inter-urban) and the intra-urban.

The first involves a system of cities or regions with a relatively coarse level of resolution: each village, town or city within it is likely to be represented by one zone only; and indeed, there may be a system of regions, one region to a zone, where each region contains several urban settlements. In the intra-urban case, the focus is on a single city or an urban region, and the zone system is defined so that areas within the city are distinguished and questions of intra-urban structure can be addressed. Most of the argument below is conducted in terms of the second kind of largely urban system, though most of it is applicable at either scale.

A NOTE ON INTER-REGIONAL SYSTEMS

There are elements of models that can be used within inter-regional systems which are not usually applicable at the intra-urban scale. These include the multi-region demographic models in a style first developed by Leslie (1945, 1948), the 'classic' origins of which date back to Malthus (1798) and,

more recently, Volterra (1938), and the input–output economic models invented by Leontief also in the 1930s (summarised in Leontief, 1967).

In each of the demographic and economic cases, the model can be built for a single region, and the multi-regional system is modelled as a set of single-region models with flows (involving other submodels) which provide the connections between them. In the demographic case, the 'connections' take the form of migration models (which, as we have seen, have a history dating back to Ravenstein in the 1880s); in the economic case, they are trade flows. Both phenomena occur, of course, in relation to the smaller zones of an intra-urban system, but they are not usually thought of, or modelled, in the same terms (though there is an argument to be presented in Chapter 8 that these ideas need to be developed further). In the urban case, the equivalent population flows are modelled in relation to residential relocation; and the equivalent economic flows as part of the industrial location and service location sectors. However, there is one concept that it is useful to carry over from the inter-regional to the intra-urban system: that of *accounting*.

Both the demographic and economic models are account-based models in that they trace the 'life histories' of the relevant entities. This both facilitates model-building and also ensures a degree of self-consistency in the models. At least the second of these reasons makes some form of accounting relevant to the intra-urban scale, and it is a desirable feature in any model system.

KEY IDEA 6.1

Accounting means tracing the elements of a system – keeping track. This is critical for building consistency into models. This, in turn, is always likely to improve goodness of fit.

It is useful to show briefly what these kinds of inter-urban accounts look like in a more formal notation and this also enables us to show how a proper handling of spatial interaction can solve, in a simple way, quite common geographical problems. However, since this would deflect us from the core of the argument at this stage, these ideas are presented in Appendix 3.

Spatial interaction models

MATRIX ADJUSTMENT

We now turn to the advent of better spatial interaction models. These had their origins in the post-war period with increasing car ownership and urban sprawl in the United States. By the mid-1950s major studies of transport patterns were under way in a number of American cities, the most important ones being in Detroit (perhaps appropriately) and Chicago, both under the direction of J. D. Carroll (see Carroll, 1955). The planning pressure for these studies coincided with the advent of large computers, and a new style of mathematical model was practically feasible for the first time.

The first attempts to model transport flows involved carrying out a large survey of origin–destination flows and then factoring the resulting matrix to generate a forecast. We can build on the notation introduced in the introduction. Let $\{T_{ij}^{\text{obs}}\}$ be the observed matrix, and suppose that *forecast* origins and destinations were $\{O_i\}$ and $\{D_j\}$. How was it possible to obtain a forecast a transport flow matrix $\{T_{ij}\}$ such that

$$\Sigma_j T_{ij} = O_i \qquad (6.3)$$

and

$$\Sigma_i T_{ij} = D_j \qquad (6.4)$$

(See Appendix 1 and Figure A1.1 for an explanation of the notation and an intuitive account of what these equations mean.) Fratar (1954) discovered an iterative procedure for solving this kind of problem which can now be recognised as using a fixed-point theorem, though it was not then presented in that way. It can be written

$$T_{ij} = A_i B_j T_{ij}^{\text{obs}} \qquad (6.5)$$

with

$$A_i = \Sigma_k O_i / B_k T_{ik} \qquad (6.6)$$

and

$$B_j = \Sigma_k D_j / A_k T_k \qquad (6.7)$$

which then ensure that (6.3) and (6.4) are satisfied. Note that each A_i equation depends on all the B_js; and vice versa. And so the equations have to be solved iteratively. This taught the transport engineers how to introduce factors A_i and B_j into models which would achieve something more than the single K of the old gravity model (cf. equation (5.6) of the previous chapter).

> ### KEY IDEA 6.2
>
> This procedure for adjusting matrices to new row and column totals is a very powerful one. It has been used in other fields – notably input–output analysis where it was introduced by Richard Stone (1967, 1970) under the title RAS method. Underpinning it is some deep mathematics due to Brouwer (1910).

> ### KEY IDEA 6.3
>
> The transport matrix, and the emphasis on satisfying the constraint equations (6.3) and (6.4) connect to the idea emphasised earlier of *accounting*: a key model-building principle will be to account for the people in the systems and their activities – in this case, travel.

TRANSPORT AS A FUNCTION OF LAND USE: THE DOUBLY CONSTRAINED SPATIAL INTERACTION MODEL

There was an important theoretical insight, usually attributed to Mitchell (e.g. by Harris, 1964), which adds the next stage of the argument in building a transport model: that transport flows must be a function of land use; and that adequate forecasts could only be obtained from a model which made this explicit rather than by factoring an observed flow matrix. It was then a relatively short step to use factors such as the A_i and B_j in (6.5) to build a refined gravity model, now more properly called a spatial interaction model because of the different 'factors' and the more general form of distance function: instead of d^{-2} we use $c^{-\beta}$ with a more general parameter, β, and c_{ij} as a measure of travel cost. Then

$$T_{ij} = A_i B_j O_i D_j c_{ij}^{-\beta} \qquad (6.8)$$

where

$$A_i = 1/\Sigma_k B_k D_k c_{ik}^{-\beta} \qquad (6.9)$$

and

$$B_j = 1/\Sigma_k A_k O_k c_{kj}^{-\beta} \qquad (6.10)$$

to ensure that (6.3) and (6.4) are satisfied. (Equations (6.9) and (6.10) can be derived simply by substituting from (6.8) into (6.3) and (6.4) and rearranging.) The number of trips generated, origins (or productions), O_i, and destinations (or attractions), D_j, are estimated in a separate submodel as functions of land use and associated activity variables.

A THEORETICAL FRAMEWORK: THE ENTROPY-MAXIMISING MODEL

A rationale for this model was offered through entropy-maximising theory (Wilson, 1967). Anyone who has studied physics, and in particular, statistical mechanics, will recognise the factors A_i and B_j as related to the partition functions which turn up in the physics of gases. The same idea can be applied in the transport context. If it can be assumed that we are interested in *bundles* of trips (as in T_{ij} from a particular i to a particular j) and not individuals, then it can be shown that the probability occurring of a pattern of trips – essentially the matrix $\{T_{ij}\}$ – is proportional to the number of ways in which individuals could be arranged in these bundles to produce that pattern. If T is the total number of trips, combinatorial theory shows this number for a particular pattern $\{T_{ij}\}$ to be

$$S = T!/\Pi_{ij} T_{ij}! \qquad (6.11)$$

We can therefore derive a most-probable pattern by maximising this function, which in physics is an entropy function, subject to the constraints (6.3) and (6.4). However, we have to add a constraint which reflects the propensity of people to travel. If the total amount spent on travel is C, then

$$\Sigma_{ij} T_{ij} c_{ij} = C \qquad (6.12)$$

is an appropriate constraint. To simplify slightly, we next note that to maximise the entropy function, we can equivalently maximise its logarithm and hence (with some manipulation and using Stirling's approximation) write S (ignoring the constants) as

$-\Sigma_{ij} T_{ij} \log T_{ij}$. The full maximisation problem can then be written as:

$$\text{Max}_{\{T\}} S = -\Sigma_{ij} T_{ij} \log T_{ij} \qquad (6.13)$$

subject to

$$\Sigma_j T_{ij} = O_i \qquad (6.14)$$

$$\Sigma_i T_{ij} = D_j \qquad (6.15)$$

and

$$\Sigma_{ij} T_{ij} c_{ij} = C \qquad (6.16)$$

and the solution to this maximisation problem can be shown to be

$$T_{ij} = A_i B_j O_i D_j \exp(-\beta c_{ij}) \qquad (6.17)$$

where

$$A_i = 1/\Sigma_k B_k D_k \exp(-\beta c_{ik}) \qquad (6.18)$$

and

$$B_j = 1/\Sigma_k A_k O_k \exp(-\beta c_{kj}) \qquad (6.19)$$

Fortunately, it turns out that the most probable state is very much the most probable – which no doubt accounts for the success and robustness of the model's performance. It is worth emphasising that although this idea was introduced as one borrowed from physics, it in no way relies on this analogy. It is in effect a high-level supradisciplinary concept which has a wide range of application – for model-building in any Weaver-type system of disorganised complexity.

KEY IDEA 6.4

Entropy-maximising methods can be used to derive robust most-probable state models of large-population systems.

It is worth noting at this point that the constraint equations can be relaxed and replaced by inequalities, thus generalising the model. This is well described by Robinson (1998), based on the work of Pooler (1994).

There is one difference between the statement of the model in equations (6.17) to (6.19) and those derived more pragmatically as equations (6.8) to (6.10): the travel cost impedance function, which was a power function, $c_{ij}^{-\beta}$, in the former has been replaced by a negative exponential function, $\exp(-\beta c_{ij})$ in the latter. We will see shortly that this helps us with interpretation.

How can we interpret this model intuitively? The trip matrix element, T_{ij}, is *approximately* proportional to the total number of origins, O_i, and the total number of destinations, D_j. We have to say 'approximately', because the pattern is modified by the appearance of the O_i and D_j terms in the balancing factors A_i and B_j. We will see shortly that we will be able to interpret the balancing factors in terms of rents in imperfect markets. These modifications reflect the competition of alternative work destinations when viewed from i; or alternative employees when viewed from employers in j. And we add to this that the number of trips from i to j is a declining function of the travel cost between them. As c_{ij} increases, $\exp(-\beta c_{ij})$ decreases. It decreases faster the larger the β, which in turn correlates with the smaller the total expenditure on travel, C, in equation (6.16). Because the negative exponential function emerges directly from the derivation, what does this imply for the power function that is sometimes used? Note that if we replace c_{ij} in the exponential function by $\log c_{ij}$, then

$$\exp(-\beta \log c_{ij}) = \exp[\log(-\beta c_{ij})] = c_{ij}^{-\beta} \quad (6.20)$$

and an effective way to interpret this is to argue that people are behaving as though they perceive travel cost to be increasing logarithmically in relation to what is usually measured. We might expect, for example, that if trips are in general short, then people will perceive travel cost to be linear, and the exponential function will fit best. If trips are much longer for example, as in a regional system, then people may be less worried about adding further to longer trips, and the power function may fit better. This becomes a matter of empirical testing – and we can then use the modelling framework to interpret the results.

KEY IDEA 6.5

The model framework can be used to interpret aspects of human behaviour, such as how travellers perceive travel cost in different situations. If the power function fits best, then this mean that people, on average, are perceiving travel cost logarithmically.

ALTERNATIVE DERIVATIONS

The entropy-maximising approach was followed by a host of alternative derivations. As hinted in Chapter 4, it would even be possible to construct an entropy-maximising model using the concepts of neural computing (using the methods of Deco and Obradovic, 1996). However, the important key point is that a robust and useful spatial interaction model can be developed. To some extent, the interpretation adopted by a particular user is a matter of taste. However, it is probably more productive to have an eye on each of the alternatives to see what can be added to the interpretation of the model. In this respect, perhaps the most important alternative is the economic one. A full review is given in Macgill and Wilson (1979) and Wilson and Macgill (1979) (see also Wilson *et al.*, 1981). For present purposes, only two key points need to be made in the broadest terms.

To base an interaction model in neoclassical economics, it is necessary to define a utility function that shows the net benefit of making a trip. If this was optimised, then everyone in a given residential location would choose the same destination, which is unrealistic. One approach is to define an average utility and then add a random component to reflect different perceptions in the population. For an (i, j) combination, let U_{ij} be that average and let ε be a random component. We then need to calculate the probability, p_{ij}, that an individual chooses the (i, j) combination. It can be shown that

$$p_{ij} = \exp(\beta U_{ij})/\Sigma_{ij}\exp(\beta U_{ij}) \quad (6.21)$$

provided that ε has a particular distribution – a Weibull distribution (see Williams (1977) for an exposition).

If the only element of utility was a cost, so that

$$U_{ij} = -\beta c_{ij} \quad (6.22)$$

then (6.21) has some kind of family resemblance to the core model derived from the entropy-maximising procedure. By further elaboration, we can construct the model itself, but this is left to the case of the retail model below, in which context it is more interesting.

Not surprisingly, the Weibull distribution has a family resemblance to the entropy function! Since economists have to pull this out of the hat in order to derive a desired model, there is perhaps an Occam's razor argument for accepting the primacy of the entropy-maximising argument! However, there is one bonus from having access to the economic alternative: it is possible to derive a measure of consumers' surplus which can be used to evaluate transport improvements, and this is the second key point (cf. Williams *et al.*, 1990).

THE SPATIAL INTERACTION MODEL AND THE TRANSPORTATION PROBLEM OF LINEAR PROGRAMMING

Consider the following optimisation problem, using the variables already defined: what set of T_{ij} values minimise the cost of delivering D_j units to each zone j from amounts O_i available at each origin i? Formally, this can be written:

$$\text{Min}_{\{Tij\}}C = \Sigma_{ij}T_{ij}c_{ij} \qquad (6.23)$$

subject to

$$\Sigma_j T_{ij} = O_i \qquad (6.24)$$

and

$$\Sigma_i T_{ij} = D_j \qquad (6.25)$$

This is a standard linear programming problem originally formulated by Hitchcock in 1941. In the solution to this problem, there is a relatively small number of non-zero flows, i.e. non-zero T_{ij}s. In this sense, it is not a realistic alternative to the spatial interaction model derived by entropy-maximising methods. The data show most real travel matrices to have mostly non-zero elements – and certainly to be non-optimal in the linear programming sense. However, there is an interesting relationship between the models which provides an important new insight.

It is not difficult to imagine (though the proof will not be given here) that the entropy-

maximising formulation in equations (6.13)–(6.16) can be written

$$\text{Max}_{\{Tij\}}Z = -(\Sigma_{ij}T_{ij}\log T_{ij})/\beta - \Sigma_{ij}T_{ij}c_{ij} \qquad (6.26)$$

subject to (6.24) and (6.25). Now as $\beta \to \infty$, the first term in Z tends to zero, and maximising Z is equivalent to minimising C in equation (6.23). This was formally proved by Evans (1973). β, therefore, can be interpreted as measuring the importance of transport costs to travellers in a particular system. If β is very large, the trips will be very short and the pattern will be closer to that of the transportation problem of linear programming. As β becomes small, then cost is relatively unimportant and trips will be more dispersed. We will see later when we present a more general version of this argument that the linear programming solution can be considered to be a perfect market outcome and the finite-β solution to represent a (more realistic) imperfect market. These arguments will be presented later in the chapter.

ELABORATIONS AND EXTENSIONS

In this section so far, we have focused on the core interaction model within the transport model in order to concentrate on the central idea. In this subsection, we note the elaborations and extensions which the reader may want to pursue, without attempting to document the detail.

The obvious elaborations are in terms of the detail of a transport system. Different types of people (say type n) have different transport behaviour. The system itself is made up of different modes (say that k is a typical mode). Trips are for different purposes, say type p, and parameters will vary by purpose. The trips themselves are carried on modal networks (road, rail and so on), and we have to find a way of characterising the links of these networks. Care needs to be taken that the model only represents mode availability for a person type if it is available to that person type. Most obviously, non-car owners do not have the option of travelling by car!

The model can be extended in a straightforward way. Using an obvious notation, the core interaction model for multiple modes and person types (still for a single purpose) can be written

$$T_{ij}^{kn} = A_i^n B_j O_i^n D_j \exp(-\beta^n c_{ij}^k) \qquad (6.27)$$

The number of trip origins is assumed to vary by person type, but not the number of destinations; generalised travel cost varies by mode. This enables us to work out, for example, the number of people travelling by mode k. Define $\gamma(n)$ to be the set of modes available to people of type n. Equation (6.27) then only holds for $k\varepsilon\gamma(n)$. $T_{ij}^{kn} = 0$ otherwise. We can define a special summation $\Sigma_{k\varepsilon\gamma(n)}$ to mean 'summation over those modes available to people of type n'. Then

$$M_{ij}^{kn} = \exp(-\beta^n c_{ij}^k)/[\Sigma_{k\varepsilon\gamma(n)}\exp(-\beta^n c_{ij}^k)] \qquad (6.28)$$

is the proportion of mode k trips by type n people.

The transport interaction model is usually assumed to have four components which lead from one to the other:

trip generation

↓

trip distribution

↓

modal split

↓

assignment

The *trip generation* model predicts the O_is and D_js as functions of land use, and so connects the transport model to more slowly-varying structural variables. The trip distribution model is the interaction model we have been working with. The modal split model subdivides these trip bundles by mode. (In the example in equation (6.27) we have a combined distribution–modal split model.) The assignment model loads these modal bundles on to the links of a network. It is only in this final stage that an estimate can be made of some of the elements of generalised travel cost (such as travel or waiting times) and so this means, not untypically with these kinds of models, that the whole system

has to be solved iteratively: the distribution model has to be rerun with newly estimated c_{ij}^ks.

Further theoretical extensions can be added to this account. It is more difficult for example to handle the individual elements within the random utility framework, but it can be done with suitable assumptions. It is also possible to extend the mathematical programming derivation of the entropy-maximising model to deal with all elements simultaneously.

For present purposes, the reader is urged to focus on the key ideas and assume that the detail can be added as required, though it will already be clear from this brief account that considerable effort and skill are required to do this.

KEY IDEA 6.8

Once a key idea has been understood conceptually, it is not too difficult to add detail: usually, if the detail can be articulated, the model can be easily extended.

The interaction model as a location model

THE CORE MODEL

By the mid-1960s, another crucial step was taken (independently) by Huff (1964), Harris (1964) and Lakshmanan and Hansen (1965). They each used a version of the model with only one set of factors, the A_is, and applied it to shopping trips. We can use the variables introduced in the introduction which are repeated here for convenience.

Let e_i, be the per capita expenditure by residents of zone i, P_i the population of zone i, and then $e_i P_i$ is the total expenditure from i; let W_j stand for the attractiveness of zone i, and for simplicity let us assume that it is measured by the amount of retail floorspace in zone j raised to a power α, so we show it as W_j^α; and let S_{ij} be the flow of expenditure on retail goods from i to j.

The key equation, which will underpin much of what follows, can then can be written

$$S_{ij} = A_i e_i P_i W_j^\alpha \exp(-\beta c_{ij}) \qquad (6.29)$$

where

$$A_i = 1/\Sigma_k W_k^\alpha \exp(-\beta c_{ik}) \qquad (6.30)$$

to ensure

$$\Sigma_k S_{ik} = e_i P_i \qquad (6.31)$$

This model can be derived by entropy-maximising methods (or any of the alternatives) as with the transport model. However, there is one critical difference already noted, but to be re-emphasised: while the A_i balancing factor is present, the B_j one is not. This is because while a set of origin constraints have to be satisfied (equation (6.31)) – and these generate the A_is – there are no destination constraint equations, and so no B_js.

What is the significance of losing the destination balancing factor? The quantity $\Sigma_j S_{ij}$, the total flow of revenue into j, is not now forced to be equal to a predetermined quantity like D_j in equation (6.3). So we can turn that equation round and use the model to *predict* D_j:

$$D_j = \Sigma_i S_{ij} \qquad (6.32)$$

which can be written

$$D_j = \Sigma_i [e_i P_i W_j^\alpha \exp(-\beta c_{ij})/\Sigma_k W_k^\alpha \exp(-\beta c_{ik})] \qquad (6.33)$$

where we obtain (6.33) by substituting for S_{ij} using (6.29) and (6.30). *Thus the interaction model can be used as the basis for predicting a locational variable* like D_j. In effect, the interaction model is being used to sort out the effects of the competition between the W_js for the spatially-distributed revenues, $\{e_i P_i\}$.

KEY IDEA 6.9

The interaction model can be used to predict a locational variable: the amount of (retail) activity attracted to each location.

This is a deceptively powerful idea, and immediately its potential in relation to classical problems can be seen. The Wj's can be seen as markets competing for customers. By the use of the discrete zone system and the spatial interaction model, *any restrictions to monocentric systems can be forgotten*. We spell out the full implications of

this in Chapter 7. Meanwhile, we summarise the position reached as a prelude to a further extension of the argument into dynamics.

SUMMARY

It is useful to take stock at this point and to assess the impact of the modelling styles discussed above on urban and regional analysis. Into the 1970s, it was increasingly recognised that the traditional gravity model was fatally flawed and that something like the entropy-maximising or random-utility models represented a step forward. There were various applications to fields as diverse as migration flows and trade flows. A few geographers used the models to assess the impacts of shopping centre development (or in relation to residential location and housing). These various submodels were also integrated into a more comprehensive model, mostly in frameworks which followed Lowry's (1964) *A model of metropolis*. (See Wilson 1971b) for a formulation which extended the Lowry model within the spatial interaction paradigm.) The models were seen mainly as *interaction* models, rather than providing a basis for locational analysis. Within geography, they were certainly rarely contrasted to classical theory. Those who continued the classical traditions on the whole did not work with these kinds of models; the modellers, in many instances, had not been brought up in the classical tradition and were often from outside geography. None the less, by the late 1970s it was being suggested that something like an extended Lowry model might *replace* central place theory for instance (cf. Wilson, 1978). As it happens, the final piece in the jigsaw was missing, and we turn turn to this later in the chapter.

A CONCLUDING COMMENT

We have already seen that there was a spate of different derivations of spatial interaction models in the 1960s and 1970s (see Macgill and Wilson (1979) for a survey). It became clear in this process that apparently different kinds of models fitted together in a 'family' of models (cf. Wilson, 1971a). We have concentrated on two examples in this chapter to focus on the essence of the idea.

Further concepts: interpretations and performance indicators

INTRODUCTION

In this section, we construct some new concepts which enable us to interpret the interaction model more effectively and which offer key performance indicators for use in applications. We proceed in four stages. We consider terms which arise directly in the model and which can be interpreted as accessibility and consumers' surplus. Then, we return to the issue of programming formulations of the model (first introduced earlier in the chapter). Finally, we construct a set of performance indicators which can be used to 'solve' some traditional problems – such as, what is a reasonable measure of a catchment area? The argument in this section largely follows that presented in Bertuglia *et al.* (1994).

ACCESSIBILITY

It is Hansen (1959) who is credited with first introducing the idea of a measure of accessibility associated with the spatial interaction model. We can illustrate the argument by continuing to use the retail model as an example, using the variables and model given in equations (6.29)–(6.31) above. Consider

$$Q_i = \Sigma_j W_j \exp(-\beta c_{ij}) \tag{6.34}$$

It can be seen that the sum on the right-hand side can be interpreted as a measure of the accessibility for residents of zone i to retail facilities: a term in the sum is a large contributor if W_j is relatively large and if j is relatively near – so that c_{ij} is small and $\exp(-\beta c_{ij})$ is relatively large. This is a reasonable intuitive measure. A single term of the sum can be considered as a measure of the accessibility to a particular zone.

If we combine equations (6.29) and (6.30) by substituting for A_i in equation (6.29), we see that

$$S_{ij} = e_i P_i W_j^\alpha \exp(-\beta c_{ij})/\Sigma_k W_k^\alpha \exp(-\beta c_{ik}) \tag{6.35}$$

This formulation shows that the model allocates the spending power of the residents of zone i ($e_i P_i$)

among the retail centres j in proportion to their relative accessibility from zone i and so offers a useful intuitive interpretation.

Later, when we show how models can be disaggregated, it will be straightforward to disaggregate the accessibility concept if appropriate.

CONSUMERS' SURPLUS

We now return to the formulation of the model as given in equation (6.29) which we repeat here for convenience:

$$S_{ij} = A_i e_i P_i W_j^\alpha \exp(-\beta c_{ij}) \tag{6.29}$$

With some simple algebraic manipulation, this can be written in the form

$$S_{ij} = A_i e_i P_i \exp\{\beta[(\alpha/\beta)\log W_j - c_{ij}]\} \tag{6.36}$$

Since $-c_{ij}$ can be taken as the disutility of travel for someone choosing to go from i to j to shop, this formulation, by considering the term in square brackets, suggests that we can interpret $(\alpha/\beta)\log W_j$ as the benefit achieved by using a retail centre of size or attractiveness, W_j. The relative size of the parameters α and β determines the relative importance of size benefits and travel costs, and of course these will be different for different kinds of goods and services. As in the case of accessibility indices, these measures can be disaggregated as appropriate.

It then follows from this argument that

$$Z = \Sigma_{ij} S_{ij}[(\alpha/\beta)\log W_j - c_{ij}] \tag{6.37}$$

can be taken as a measure of benefit to consumers for the system as a whole. Coelho and Wilson (1976) showed that the maximisation of this function is equivalent to the maximisation of the formal concept of consumers' surplus. See also Champernowne *et al.* (1976) and Williams (1977).

PROGRAMMING FORMULATIONS, RENTS AND IMPERFECT MARKETS

We indicated above that the transportation problem of linear programming is a limiting case of the spatial interaction model. The model, as represented by equations (6.23)–(6.25), is repeated here for convenience:

$$\text{Min}_{\{T_{ij}\}} C = \Sigma_{ij} T_{ij} c_{ij} \quad (6.23)$$

subject to

$$\Sigma_j T_{ij} = O_i \quad (6.24)$$

and

$$\Sigma_i T_{ij} = D_j \quad (6.25)$$

We can now extend the formal argument by noting the fact that any programming problem has a dual formulation. Thus, if equations (6.23)–(6.25) are taken as the primal problem, the dual is written by introducing variables associated with the constraints and formulating an appropriate objective function. Let $\{\mu_i\}$ be a set of variables associated with the constraints (6.24) and $\{\nu_j\}$ a set associated with equation (6.25). Then the new and equivalent objective function for the dual formulation is

$$\text{Max } C' = \Sigma_i \mu_i O_i + \Sigma_j \nu_j D_j \quad (6.38)$$

subject to

$$c_{ij} - \mu_i - \nu_j \geqslant 0 \quad (6.39)$$

μ_i and ν_j can be interpreted as measures of comparative advantage (or, again, benefit) and in terms of economic theory, therefore as the rents which can be extracted by landowners because of this comparative advantage.

It was argued earlier in the chapter that the entropy-maximising formulation of the interaction model is a nonlinear programming problem (because of the introduction of the entropy term into the objective function) and that the *linear*

transportation problem of linear programming is a special case. However, nonlinear programming problems as well as linear ones have dual formulations. We can again introduce dual variables μ_i and ν_j associated with the constraint equations and it turns out in the nonlinear case that the dual has an analytical solution:

$$T_{ij} = \exp(-\mu_i - \nu_j - c_{ij}) \quad (6.40)$$

μ_i and ν_j can still be interpreted as measures of comparative advantage, or rents, this time in an imperfect market. They will be smaller than their values in the equivalent linear problem. The amount of imperfection is determined by the parameter β: the smaller it is, the greater the degree of imperfection. It can now also be seen that the balancing factors A_i and B_j are transformations of the μ_i and ν_j, and so can themselves be interpreted in terms of comparative advantage and rents in imperfect markets. In particular:

$$A_i O_i = \exp(-\mu_i) \quad (6.41)$$

and

$$B_j D_j = \exp(\nu_j) \quad (6.42)$$

This argument can of course be extended to the singly constrained model, with A_i being interpreted as a comparative advantage and related to a rent. See Wilson and Senior (1974) for the original argument and for a detailed exposition.

KEY IDEA 6.12

The duals of the programming formulations of spatial interaction models show that key terms can be interpreted as comparative advantage or rents.

CATCHMENTS AND PERFORMANCE INDICATORS

We now return to the retail model as the best example to fix ideas for the introduction of the next set of concepts. It will be argued in this section that concepts can be derived from the spatial interaction model which solve an old problem and add an important new concept, and which can then be used as the basis for constructing performance indicators related to various aspects of the system of interest. It will be easy to see how the ideas developed in the retailing concept can be applied more broadly.

The 'old problem' is that of a catchment area. Conventionally, attempts are made to draw boundaries around retail centres and the population within the boundary summed to obtain what is called the *catchment population*. The problem is that this is often done by different people without consistent boundary definitions, and there tends to be a lot of double counting. The root of the issue, of course, as the interaction models indicate, is that flows always overlap any boundaries that are drawn. A new kind of measure is needed – one which has the property that when the catchment populations of the retail centres in a region are summed, they add up to the total population of the region.

We use the notation that an asterisk replacing a subscript denotes summation, so that

$$S_{i*} = \Sigma_j S_{ij} \qquad (6.43)$$

is the total of trips (or cash) leaving i. Then S_{ij}/S_{i*} is the proportion of trips (cash) going from i to j. A measure of the catchment population of the retail centre at j is then:

$$\Pi_j = \Sigma_i (S_{ij}/S_{i*})P_i \qquad (6.44)$$

Note that, with this definition,

$$\Sigma_j \Pi_j = \Sigma_i P_i \qquad (6.45)$$

It is then possible to calculate indicators, say for the management of a retail centre, such as D_j/Π_j, with a meaningful denominator.

The new concept is obtained by using the mirror image of this argument. Consider:

$$\Omega_i = \Sigma (S_{ij}/S_{*j})W_j \qquad (6.46)$$

This allocates a proportion of W_j to residential zone i, the proportion being determined by the proportion of j's users who come from i. Ω_i can be interpreted as the volume of retail facilities that are *delivered* to the population at zone i. An indicator such as Ω_i/P_i then measures the relative provision at i.

We are always seeking to achieve both efficiency and effectiveness in systems. An indicator such as D_j/Π_j can be regarded as an *efficiency* indicator while Ω_i/P_i is an *effectiveness* one. It is possible for centres to be efficient without

the system being effective, in that there may be 'pockets' of relative deprivation – zones which have low delivery indicators. And it is possible to be effective without being efficient! The Hotelling problem introduced in Chapter 5 and detailed in Appendix 2 also illustrates very clearly that the consumer optimum location pattern for retail centres, and the retailer optimum, may be very different. See Clarke and Wilson (1987a, b) for a detailed presentation of the associated arguments. We return to these topics via associated research issues in Chapter 8.

Structural dynamics

THE KEY HYPOTHESIS

To fix ideas and for simplicity, we will continue the presentation in terms of the shopping model and all the relevant variables defined earlier. We begin by discussing how the shopping interaction model could be and was used in locational analysis and then see how this offers the seed for an extension towards the crucial hypothesis and model which allows geographical structure to be modelled in a wide variety of circumstances.

Suppose we are a group of planners using the shopping model given by equations (5.29) and (5.30). We produce a trial plan $\{W_1, W_2, \ldots W_N\} = \{W_j\}$. These values are input to the model. We assess the impact of this plan by calculating

$$D_j = \Sigma_i S_{ij} \qquad (6.47)$$

from the model. For ease of argument, let us assume that the cost, C_j, of supplying and running facility W_j at j is a linear function, though with costs possibly varying by zone:

$$C_j = k_j W_j \qquad (6.48)$$

Then since D_j is a measure of revenue, and assuming compatibility of units, if $D_j > C_j$, then j is a good site, a profitable centre, or whatever, and it could in principle be expanded – and vice versa. This model 'suggests' likely directions of change for a new round of planning.

This is often a useful mode of model use, but we can now take a crucial further step (Harris and Wilson, 1978): we can make an assumption and formalise the hypothesis implicit in the above account of a planning procedure, now assuming that we are going to model the market for retail floorspace supply – the first step towards modelling what we have hitherto taken as a fixed (or slowly varying) structural set of variables. The assumption is that market forces will bring about a spatial structure, $\{W_j\}$, in relation to profits: it is assumed that developers and retailers will behave in this way. So the more formal hypothesis is: if $D_j > C_j$, let W_j expand by an amount proportional to $D_j - C_j$; and vice versa. If we assume that the change takes place in some period, say a year, this can be written as a difference equation:

$$\Delta W_j(t, t+1) = \varepsilon[D_j(t) - C_j(t)] \tag{6.49}$$

for a suitable parameter, ε. The equilibrium condition is

$$D_j = C_j \tag{6.50}$$

It can now be argued that these relatively simple equations, worked out here for a particular example, provide the basis of the modelling part of the theoretical revolution we have been seeking. To see that the simplicity is deceptive, we first write out the model in full. Repeating (6.29) and (6.30) for convenience, the interaction model is as follows:

$$S_{ij} = A_i e_i P_i W_j^\alpha \exp(-\beta c_{ij}) \tag{6.29}$$

with

$$A_i = 1/\Sigma_k W_k^\alpha \exp(-\beta c_{ik}) \tag{6.30}$$

We can also take as a cost assumption

$$C_j = k_j W_j \tag{6.48}$$

(repeating equation (6.48)) if we assume linearity, or, we can write a more general functional relationship:

$$C_j = C_j(W_j) \tag{6.51}$$

Then, substituting for D_j and C_j in (6.49) and (6.50), we get

$$\Delta W_j = \varepsilon\{\Sigma_i[e_i P_i W_j^\alpha \exp(-\beta c_{ij})/\Sigma_k W_k^\alpha \exp(-\beta c_{ik})] - k_j W_j\} \tag{6.52}$$

for the difference equations and

$$\Sigma_i[e_i P_i W_j^\alpha \exp(-\beta c_{ij})/\Sigma_k W_k^\alpha \exp(-\beta c_{ik})] = k_j W_j \tag{6.53}$$

for the equilibrium conditions (where we have dropped the t and $t+1$ labels for simplicity).

Then, either we can calculate $\{W_j\}$ from (6.52) for successive time periods, with

$$W_j(t+1) = W_j(t) + \Delta W_j(t, t+1) \tag{6.54}$$

at each step (if we do this, then, other things being equal, the system will be approaching equilibrium but will not actually be in equilibrium); or we can assume that the system *is* in equilibrium and solve (6.53) directly for $\{W_j\}$. The next step in the argument is to explore the equilibrium solution and then we can move on to the dynamics.

KEY IDEA 6.14

By making an assumption about the cost of retail infrastructure (which is a component of retail attractiveness), and an assumption about the behaviour of retailers, it is possible to write down the equations for, and hence begin to model, the equilibrium and dynamics of structural variables. It will turn out to be possible to extend this idea very widely.

THE EQUILIBRIUM SOLUTION: GEOGRAPHICAL STRUCTURE

For the time being, let us concentrate on the equilibrium solution. The first striking feature of the equations is that there is no means by which we can find an analytical solution for $\{W_j\}$. To make it clear just how messy the equations are, we can write the left-hand side of equation (6.53) without the summation signs for a typical j:

$$\begin{aligned} &e_1 P_1 W_j^\alpha \exp(-\beta c_{1j})/\Sigma_k W_k^\alpha \exp(-\beta c_{1k}) \\ &+ e_2 P_2 W_j^\alpha \exp(-\beta c_{j1})/\Sigma_k W_k^\alpha \exp(-\beta c_{2k}) \\ &+ e_3 P_3 W_j^\alpha \exp(-\beta c_{j1})/\Sigma_k W_k^\alpha \exp(-\beta c_{3k}) \\ &+ \ldots = k_j W_j \end{aligned} \tag{6.55}$$

There are problems caused by the exponent, α; but even with $\alpha = 1$, the main complications are caused by the denominators of each of the terms on the left-hand side of equation (5.52). Consider one of these terms, say the first, written as

$$e_1 P_1 W_j^\alpha \exp(-\beta c_{1j}) / [W_1^\alpha \exp(-\beta c_{11})$$
$$+ W_2^\alpha \exp(-\beta c_{12}) + \ldots] \qquad (6.56)$$

The denominator is made up of all the terms of the same form as in the numerator. The whole term can be interpreted as follows. The amount of the expenditure $e_1 P_1$ from zone 1 which is allocated by the model to zone j is proportional to

$$W_j^\alpha \exp(-\beta c_{1j})$$

which is a measure of the attractiveness of the retail centre in zone j *as perceived from zone 1.* That is, it is partly determined by the size component, W_j^α, and partly by nearness or access as measured by $\exp(-\beta c_{1j})$. The denominator can now be seen as the sum of all such terms for all other zones, so that the expression multiplying into $e_1 P_1$ in equation (6.55) can be interpreted as the *relative* attractiveness of the shopping centre in zone j, with the relativeness being determined by the sizes and locations of (from 1's point of view) the *competition*. If $\exp(-\beta c_{1j})$ is large compared to $\exp(-\beta c_{ij})$, $i \neq 1$, then the flow will be substantial. If it is small, the flow will be small. The model handles competitive processes in a way in which the classical models do not, hence facilitating a comprehensive approach. Thus a proportion of $e_1 P_1$ is allocated to zone j as determined by this mechanism. The next term in equation (6.55) allocates a proportion of $e_2 P_2$ to zone j, the next of $e_3 P_3$ and so on. If large amounts are being allocated from zone 1 to zone j, then j is a potentially attractive site for consumers resident in zone 1.

The complications of α and the denominator terms mean that the equations are, in mathematical parlance, *nonlinear.* The equations (6.53) (or (6.55)) are *nonlinear* simultaneous equations. Because each j equation in (6.53) contains all the W_k, the equations also exhibit a high degree of *interdependence* – in geographical terms as a consequence of trying to resolve the issues of spatial competition. We have also already noted that in these circumstances, it is impossible to obtain an analytical formula for each W_j. However,

the equations can be solved iteratively: start with a trial vector $\{W_k\}$ and successively adjust them until the equations are solved. (One practical way to do this is to run the dynamical model, (6.52), with a low ε, for many time periods, and that will converge to the equilibrium solution.)

A valuable way to explore the properties of this kind of system is to solve the equations for a range of values of the parameters and for a hypothetical grid system. This generates the kind of results shown in Figure 6.1 (from Clarke and Wilson, 1985b). There are very striking changes in pattern: for high α and low β, there are fewer larger centres. This fits (as it obviously should!) with interpretations of α and β: α measures the importance of size to consumers, and β the ease of travel (it may be lower with increasing car ownership, for example).

Even though results have only been presented for an idealised system, the potential of the approach for modelling geographical structures, and more generally complex spatial systems, should by now be clear. However, the example used illustrates one further important and dramatic feature: it turns out that there are *critical values* of parameters like α and β at which the nature of the spatial pattern changes. In Figure 6.1, for example, the centres which 'disappear' as α increases or β decreases do not decline continuously to zero; at some critical value, they suddenly vanish. This is a general property of nonlinear interdependent systems and has important applications in the modelling of changing geographical structures. It is almost certainly the case, for example, that the relatively sudden transition from a dispersed 'corner-shop' food retailing system to a concentrated 'supermarket' one can be described in this way (cf. Poston and Wilson, 1977; Wilson and Oulton, 1983).

Although we cannot solve the equations for $\{W_j\}$ analytically, we can achieve some insight by approaching equation (6.53) graphically. Consider the left-hand side of (6.53), which is D_j, as a function of W_j, say $D_j(W_j)$:

$$D_j(W_j) = \Sigma_i [e_i P_i W_j^\alpha \exp(-\beta c_{ij}) / \Sigma_k W_k^\alpha \exp(-\beta c_{ik})] \qquad (6.57)$$

The right-hand side, $k_j W_j$ is the cost, C_j, and this, if plotted against W_j, is a straight line. We can then

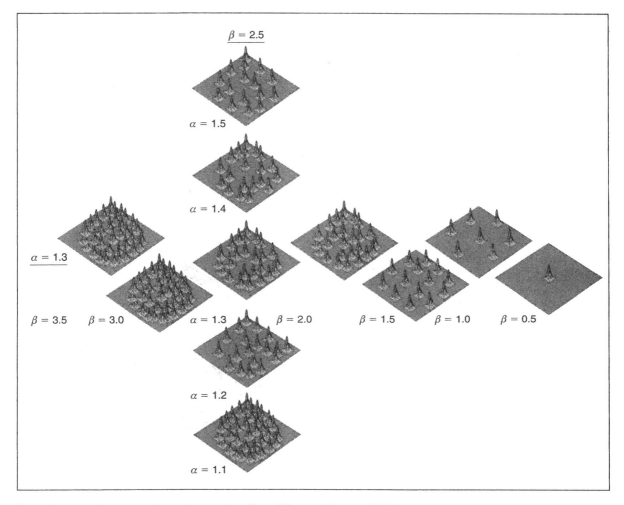

Figure 6.1 Spatial structures for varying α and β. (From Wilson and Bennett, 1985)

explore the functional form of $D_j(W_j)$ when plotted against W_j. It turns out that these forms vary with α. For $\alpha < 1$, the gradient is infinite at the origin; for $\alpha = 1$, it is finite; and for $\alpha > 1$, it is zero. The different cases are shown in Figure 6.2. For $\alpha = 1$, there are two variants; and for $\alpha > 1$, there are three. These arise when the cost straight line is added to the figure. The point of interest is whether the line intersects the curve at a non-zero value or not.

The first point to note is that the origin, $(W_j, D_j) = (0, 0)$, is always a solution. In the $\alpha < 1$ case, in addition there is always a finite solution for each W_j. We then assume that this will be the case in reality: some entrepreneurs will identify the

opportunity and establish a centre. Hence the $\alpha < 1$ situation represents a dispersed system.

For $\alpha \geq 1$, however, there are situations where there is not always a non-zero solution (Figure 6.2b(ii), c(ii) and c(iii)) either because k_j is too high (and the line is too steep) or α is too low (and the revenue curve does not 'pick up' rapidly enough). In these cases, therefore, typically, there will be a spatial pattern in which there will be fewer than N centres (in an N-zone system). (The shape of the revenue curve in different situations is also dependent on β of course.) See Clarke and Wilson (1985b) for a detailed presentation of these results.

For $\alpha < 1$, there is almost certainly a unique equilibrium solution. For $\alpha \geq 1$, there will be

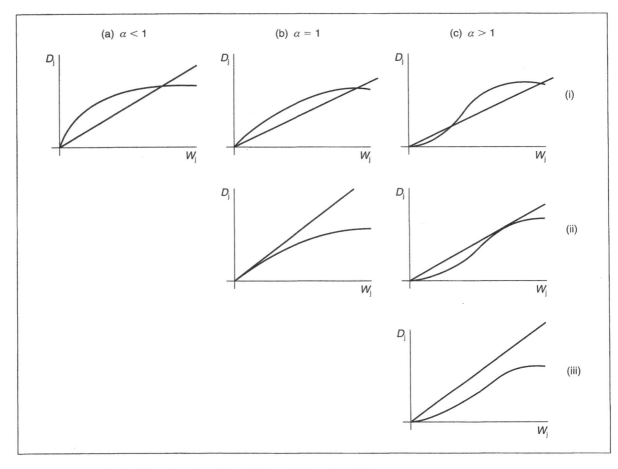

Figure 6.2 Criticality for different α values. (From Wilson and Bennett, 1985)

multiple solutions. It will be possible for a particular (α, β) combination to generate a *type* of spatial pattern, but the particular solution which occurs will depend on the initial conditions, and in effect the accidents of history as particular agents make particular decisions at particular times. It is at this point in the argument that we can see some similarities with Arthur's (1994a) work. $\alpha > 1$ in the W_j^α term does seem to represent what he calls increasing returns to scale and we have seen that it is this feature which generates multiple solutions and what he calls *path dependence* in urban evolution. However, his own equations have a simpler form than those used here and do not capture the real complexity of the system. In his model of industrial location, for example, he has two terms in an equation which represent the

probability of the Nth firm (which may, in our context be a retailer) locating at j – one focused on geographical advantage and one on increasing returns. It remains an interesting research question as to whether his basic modelling idea, representing urban evolution as a stochastic process, can be combined with the basic geography of the models used here, and we return to this issue in Chapter 8.

KEY IDEA 6.15

The spatial patterns which appear in cities and regions can be explained (at least in broad terms) through the interdependence of travel impedance and centre attractiveness in consumer behaviour. This is a nonlinear system which can be modelled.

DYNAMICS: INCREMENTS, OSCILLATIONS OR CHAOS?

The next step in the argument is to review dynamical behaviour. Equation (6.49) is repeated here for convenience:

$$\Delta W_j(t, t + 1) = \varepsilon[D_j(t) - C_j(t)] \qquad (6.58)$$

In simulations, $W_j(t)$ would grow with a finite gradient at the origin as indicated in Figure 6.3a. A more common assumption is logistic growth, plotted in Figure 6.3b with zero growth at the origin and a slower pick-up. This can be achieved algebraically by adding a factor W_j to the right-hand side of equation (6.58) which changes the low-W_j dynamics but not the equilibrium position:

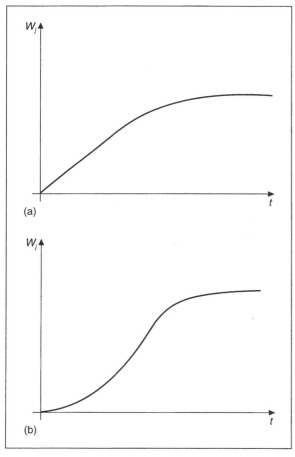

(a)

(b)

Figure 6.3 Alternative growth trajectories for a shipping centre near the origin. (From Wilson and Bennett, 1985)

$$\Delta W_j(t, t + 1) = \varepsilon[D_j(t) - C_j(t)]W_j \qquad (6.59)$$

If we substitute for C_j (as k_jW_j) and write $\Delta W_j(t, t + 1)$ as $W_j(t + 1) - W_j(t)$, then with a little manipulation, equation (6.59) can be written:

$$W_j(t + 1) = [1 + \varepsilon D_j(t)]W_j(t) - \varepsilon k_jW_j^2 \qquad (6.60)$$

and this is an interesting form because it can be compared to the standard logistic equation

$$X(t + 1) = NX(t)[1 - bX(t)/N] \qquad (6.61)$$

for some variable X and parameters N and b. May (1971, 1973) showed that for this equation, if

$$1 < N < 3 \qquad (6.62)$$

then $X(t)$ tends to a stable equilibrium. For

$$3 < N < 3.8495 \qquad (6.63)$$

$X(t)$ oscillates. And for

$$3.34895 < N < 4 \qquad (6.64)$$

chaotic oscillations set in, while for $N > 4$, there is divergence to $-\infty$, which we interpret as a collapse to $W_j = 0$. The absolute values of N are obviously important. A comparison with equation (6.60) shows that for our archetypal example

$$N = 1 + \varepsilon D_j(t) \qquad (6.65)$$

This shows that N is a function of $D_j(t)$ as well as the parameter ε.

Oscillatory behaviour implies, in this case, some sort of regular periodic frequency. Chaotic behaviour, on the other hand, means oscillations which cannot be interpreted in such a way: there is no regularity. It is the possibility of this kind of chaos occurring in natural or social systems that has attracted a lot of popular attention. Are cities susceptible to this? In practice, experience suggests that incremental change is more likely and that ε vales will be such that $1 + \varepsilon D_j(t)$ will always be less than 3. However, this initial analysis of urban and regional dynamics, based on the retail model, assumes that adjustments are made through the scale of provision in a centre – the spatial structure. There can also be other adjustments, in prices and rents, and we consider these next.

EXTENSIONS TO INCLUDE PRICES AND RENTS

In the previous section, we concentrated on the adjustment of facility sizes, $\{W_j\}$, and the calculation of an equilibrium structure, in a dynamic process which might be considered to represent the behaviour of retailers and property developers. From this standpoint, considerable insight can be gained into the structure and development of geographical systems. As we have just noted in concluding the previous section, there are at least two other possible modes of adjustment which should be considered: first, in the spatial variation in prices, say p_j as an appropriate index, for goods sold at j; and secondly, in relation to land rents, say r_j, at each zone j. Here we offer a different, and complementary, approach to rents to that of the previous section. It remains an interesting research problem to reconcile the two approaches.

Adjustment with respect to prices, $\{p_j\}$, as well as the spatial structure variables, $\{W_j\}$, is essential to establish a correct economic basis for the model. And taking account of rents is not only another extension in the direction of economic reality, but also lies at the heart of *geographical* theory: it relates to the competition of different activities for land. So we can define a vector, $\{r_j\}$, of land rents. It is important for all activities, but particularly for the dispersed ones, agricultural and residential uses, which are large consumers of land. However, one should not underestimate the importance in relation to 'point' activities either. Landowners may be able to command huge rents for small sites if they are particularly suitable, say, for superstore development, particularly if the number of such sites is restricted, e.g. by planning or zoning policies or regulations.

It is straightforward in principle to build prices and rents into the model and then to add adjustment equations. The p_j values would probably appear (on some average basis; see Wilson, 1985a, b) in the e_i-terms, to make demand elastic to price. They would certainly appear, say as $p_j^{-\gamma}$, as a factor in the attractiveness term, W_j: attractiveness would increase as prices were lowered (though there are exceptions to this which will be considered below), and hence the parameter γ is assumed positive (and $p_j^{-\gamma}$ a negative power function).

The rent would appear as an additive component in the cost C_j, which might become:

$$C_j = (k + r_j)W_j \tag{6.66}$$

where, for these illustrative purposes, the j-variation in C_j is now attributed to the rent. The adjustment equations can now be taken as

$$\Delta W_j = \varepsilon(D_j - C_j)W_j \tag{6.67}$$

dropping the t, $t + 1$ labels from equation (6.59),

$$\Delta p_j = -\mu(D_j - C_j)p_j \tag{6.68}$$

and

$$\Delta r_j = \nu(D_j - C_j)r_j \tag{6.69}$$

where we have used the logistic form of growth in each case; and μ and ν are suitable constants.

We have shown the price adjustment equations with a minus sign: success equals a reduction in prices. However, as noted earlier, we should bear in mind that there may be sectors where the opposite is the case. A very successful and profitable superstore might use the minus sign as assumed here and reduce prices in order to increase trade and profit levels in that way. A successful restaurant, on the other hand, might increase prices to control numbers and to increase profits.

In the case of land rents, the hypothesis is what Wilson and Birkin (1987) argued was an *active* one: it is specified as a function of profit, but the amount extracted as rent is, in effect, a function of the relative power of landowners and producers of crops, goods or services. The values of ε, μ and ν

then represent the relative strengths of the different agents in the various adjustment processes. The *passive* case for rent means that all adjustments are made to land use and prices of goods, and then any residual profit is assumed to be collected as rent by landowners. The von Thunen model has an implied rent hypothesis of this type.

This argument leads to another useful general point. By articulating the processes of dynamical adjustment in this way, the models become much more explicitly connected to hypotheses about the different kinds of agents involved. This provides the basis for connecting modelling to different styles of theorising more closely than has been the case in the past.

The picture is further complicated by the fact that a particular set of hypotheses can often be assembled as a model using different (albeit ultimately equivalent) mathematical methods. The most striking example is provided by the kind of model specifications used in Chapter 5, coupled with iterative solution of the equations, and alternative mathematical programming formulations of the same problem (cf. Boyce, 1984, quoting Kuhn, for a discussion of the relationships of these and other 'equivalent' mathematical formulations; also see Boyce, 1978; Wilson and Macgill, 1979).

This means that some models that look very different are actually equivalent. But there are many real differences representing alternative specifications of hypotheses which will only ultimately be resolved in the light of an extensive programme of empirical model testing.

KEY IDEA 6.17

The equilibrium and dynamic models can be extended to incorporate the spatial dynamics of the prices of consumer goods and land rents.

IMPERFECT MARKETS AND GEOGRAPHICAL STRUCTURE

We have seen (in the archetypal example used) that geographical structure is represented by the vector $\{W_j\}$. We have shown that an iterative scheme can be derived for calculating equilibrium values of $\{W_j\}$ and that difference equations can be constructed for representing the dynamical behaviour of $\{W_j\}$. In this subsection, we will show, by connecting to other ideas presented earlier, that an alternative and possibly more realistic model can be generated which stimulates thinking for further research.

The ideas were presented in an unpublished paper (Wilson, 1985a) and there remains scope for further development. The key idea is this: the analysis of equilibria and dynamics turns on the condition in each zone j that revenues balance costs: $D_j = C_j$. It can be argued that, relative to real world behaviour, this hypothesis is 'too' optimal and that the procedures used are analogous to the transportation problem of linear programming compared to the entropy-maximising method. So we can take the entropy-maximising theory and linear programming ideas from earlier in the chapter, as applied to a trip matrix $\{T_{ij}\}$, or analogously, the matrix $\{S_{ij}\}$ of this chapter, and apply them to the vector $\{W_j\}$. In that analogue, the transportation problem of linear programming produces an optimal (cost-minimising) $\{T_{ij}\}$ while the entropy-maximising method generates the most-probable sub-optimal solution. The analogous procedure for $\{W_j\}$ would therefore be the following: (i) relax the condition $D_j = C_j$ and replace it by $D_j < C_j$ – indicating that developers and retailers will find it very difficult to optimise W_j and hence C_j; (ii) add an entropy term, $-\Sigma_j W_j \log W_j$, into the equivalent mathematical programming problem and derive a new equilibrium solution. This then has the added bonus that we can derive an analytical solution for the equilibrium $\{W_j\}$. We do not pursue the details here, but the outcome can be recorded as follows:

$$W_j = \exp[\lambda(D_j - C_j)]\exp[\alpha D_j/W_j] \quad (6.70)$$

and, of course, the spatial interaction model is itself modified (see also Birkin and Wilson, 1989).

KEY IDEA 6.18

The models can be further extended to incorporate imperfect market behaviour for retailers and developers.

New theoretical foundations: a summary

To summarise, we can give our archetypal retail sales' model a more general interpretation. We can look at each of the main elements in turn as follows:

(i) E_i: there is a spatially distributed demand or need for a good or service.

(ii) W_j, X_j, F_j: these 'demands' are met in zone j and are 'pulled in' by a measure of attractiveness, W_j. In general, this will be a composite measure made up of a number of factors (say $X_1^{\alpha}, X_2^{\alpha'}, X_3^{\alpha''}...$), one of which will be the size of the facility (e.g. in terms of floor space) at j, say F_j.

(iii) S_{ij}: a spatial flow, estimated by an interaction model, not necessarily of the kind used as examples above. This will allocate each E_i among the W_js.

(iv) c_{ij}: this will be a measure of impedance between i and j.

(v) D_j: S_{ij} provides the basis for calculating the total usage attracted to j to be calculated. This may be a revenue; it may be some other measure, such as trips.

(vi) C_j: this is the cost of supplying and running the facility of size F_j at j, and so formally can be written $C_j(F_j)$. This function is, in effect, the 'production function' of the organisations offering F_j at j. There may also be costs associated directly or indirectly with some of the other factors, X_j, associated with W_j. Some of these costs may be functions of the spatial location of j relative to its suppliers. Thus, in the way that D_j is made up of a sum of spatial interaction flows, C_j may be made up of a set of such flows representing the costs of inputs to the production process.

Then it should be possible to determine $\{F_j\}$ and possibly other elements of $\{W_j\}$ by a hypothesis either of the equilibrium type

$$D_j = C_j \tag{6.71}$$

or a full dynamic treatment involving

$$\Delta F_j = \varepsilon(D_j - C_j) \tag{6.72}$$

(vii) The dynamical adjustment hypothesis can be extended by introducing a spatial distribution of the prices of good and land rent.

(viii) The model can be further extended by relaxing the constraint (6.71) and enabling the F_j to be determined in an imperfect market.

(ix) At any of these stages, the model can be disaggregated as appropriate. For simplicity, in this chapter, this level of detail has not been pursued (see e.g. Wilson, 1983a; Wilson and Bennett, 1985).

It is also worth bearing in mind – at least for fun, and possibly more productively – that this model may be generalisable beyond urban and regional analysis. Holland (1998), for instance, gives an example of a 5×5 board on which transition mechanisms can be specified which generate structures that have persistence over time. His intention is to demonstrate that emergent behaviour can be generated by simple mechanisms, and he may have had in mind the more complex versions of such persistence, as in memory, for example. However, his example is reminiscent of some locational structures in urban and regional analysis, and it turns out that his 'mechanisms' can be presented in the format of the general model presented above. (His example does show, in passing, how the structures are very much dependent on the initial conditions – the point made generally by Barrow (1991) and referred to in Chapter 4.)

> **KEY IDEA 6.19**
>
> The spatial interaction model and the associated dynamic location model provide a general method for tackling a wide range of problems. The models used as illustrations can be generalised in many ways: for example, a double interaction can be modelled – a flow from a factory (at i) to a port (at j) to a final destination (at k) to be modelled as T_{ijk}.

Concluding comments

The analysis presented means that the *interaction and location* approach is potentially applicable to a wide class of problems in urban and regional analysis, and certainly includes all the problems of classical geographical theory of Chapter 5. It can also be seen that it represents a general approach to a class of *complex spatial systems*. What is needed in each case is to find a way of setting up the problem in the format implied by steps (i)–(viii) above. This often involves ingenuity and imagination, but once the principles of the method have been understood, it can usually be achieved. It should also be emphasised that the methods are applicable at either inter-regional or intra-urban scales.

There are a number of general points to be noted:

(i) The way the discrete-zone system is used means that all the restrictions of classical theory of the 'single market centre' type disappear.

(ii) There is the added interest of being able to apply the ideas of nonlinear dynamical system theory on rapid change, and other types of bifurcation.

(iii) A framework is provided even for more conventional verbal or diagrammatic theorising, whether or not coupled with statistical hypothesis testing. The steps set out at the beginning of this section can be followed in this case as well. Indeed, they have to be even for model-building purposes. And, of course, it follows as a corollary of this argument that much of the verbal theorising and statistical presentations of this type, which still dominate much of the geographical literature, can often be improved; but can also be used as the basis for model-building. So the framework offers a bridge between traditional systematic and model-based approaches to geography.

We conclude this section by developing the second point made above because the general consequences for theory in urban and regional analysis are very great; they are of a general kind, depending only on the known presence of nonlinearities, and are independent of the particular model formulation (neoclassical or marxian for instance); and the ideas can be explained relatively easily.

The most important characteristic of nonlinear systems (and most urban and regional systems will be nonlinear in a complicated way at the very least because of spatial competition, externalities and scale economies) is that, for a given set of parameter values, they have multiple equilibrium solutions. There is usually one of these that is potentially the most stable and is known as the global equilibrium. However, in any particular real instance, when a system is in an equilibrium state (or approaching one) there is no need for it to be in the global equilibrium state. It may be in a state of a certain 'type', but one of many possible states. This accords with our intuition and experience: we may recognise certain general features of urban structure, for example, but we know, as traditional regional geographers always told us, that each city is in important respects unique. We now have a mathematical formulation for urban and regional theory – and, more generally, for complex spatial systems – which unites all these viewpoints. A historical 'accident', the siting of a particular facility at a particular place, will often partly determine the specific path of development of that system. In mathematical terms, this means that the particular equilibrium state that is reached is a function of the starting value, or of any 'disturbances'. In geographical terms, we can offer interpretations and select from the variety of possible states which the models offer in different circumstances.

Indeed, it is worth emphasising that the presentation here is such that only the flavour of the rich variety of models that can be generated has been presented. It remains a major research task to match this vast range of possibilities against empirical realities. It should also be emphasised that there are substantial mathematical problems still to be fully solved. For example, the fact that the equilibrium analysis that can be presented (e.g. in and around Figure 6.2) which shows that the analysis for any one W_j is dependent on the configuration of all the others means that we can be confident only that we have generated insights – not that we have solved the whole problem. (This particular problem has been articulated as one of *configurational analysis*; cf. Wilson, 1988).

In the next chapter, we show how in principle the ideas can be applied in a wider context; we sketch the full range of application and show how approaches to all the problems of classical theory can be transformed. Even approaches to verbal theorising can be much better informed if these ideas have been absorbed.

7

The rewriting of classical theory

Introduction

The aim in this chapter is to take the classical models of Chapter 5 and show how they can be rewritten using the methods presented in Chapter 6 – a process begun in Wilson (1989), but worked out more fully here. The results are interesting in themselves; but the narrative is more important as an illustration of model-building in the interaction–location paradigm. We consider in turn agricultural location, industrial location, residential location and housing, services, transport, and finally an integrated approach, within which we can review the task of building an integrated model. Central place theory will be considered as part of the *services* section. In virtually every case, we will see that significant progress can be made by introducing a discrete spatial representation and an associated notation. It is also interesting that in each case, there is a new generation of 'classical' authors who achieve this shift but then represent their models in an optimising mathematical programming framework. We describe these developments briefly in each section as a stepping stone to the more powerful interaction-based dynamic models.

Agricultural location

The essence of the approach will be to extend the notation introduced in Chapter 5 and Appendix 2 to represent von Thunen's model – in the first instance to switch from a continuous space representation to a discrete zone one. We will then find that we are in a position to construct a variety

of models from one of which we can derive von Thunen's model as a special case. A good notation is an important part of the model-building exercise. We will follow, in broad terms, the steps used in the summary of the previous chapter.

Assume a discrete zone system with zones labelled $j = 1, 2, 3, \ldots$ and a set of markets labelled $i = 1, 2, 3, \ldots$ (which can be considered to be the same set of zones – but more appropriately in this case, a distinct set of points representing possible market locations). We need to represent different agricultural land uses, either as crops or, for example, grazing. We will assume that the superscript $g = 1, 2, 3, \ldots$ can be used for this purpose. The system can then be characterised as follows.

Let E_i^g be the demand for the 'good' produced by use g at market centre i. Let Z_j^g be the amount of g produced in zone j. Define q^g as the amount of land needed to produce a unit of g. Another per-unit-land-area variable is rent, which we take to be r_j. Transport costs will obviously be important, and we can define c_{ij}^g as the unit cost of transporting a unit of g from j to i.

Take Y_{ij}^g as the main interaction variable: the amount of the demand for good g at i which is met from j. In the usual way, we can then define the total revenue accruing to farmers in j as D_j^g, and the total cost of producing the corresponding amount of goods, Z_j^g, as C_j^g. Profit can be defined as

$$\Pi_j^g = D_j^g - C_j^g \qquad (7.1)$$

p_i^g is the market price at i for g; v_j^g is the unit production cost at j for g. Land will obviously play an important role, and we take L_j as the total land area in j and l_j^g as the amount used for the production of good g in j.

The original von Thunen model can be characterised as having the following restrictive assumptions (described as such in a paper by Stevens (1968), who was seeking to relax them): (i) uniform fertility of the land; (ii) uniform transport costs; (iii) uniform production costs; (iv) infinitely elastic demand at a given price; and (v) a single market centre. An immediate advantage of the notation – the system characterisation – we have introduced is that we can see how to relax these assumptions. The fertility assumption could be dealt with by making q^g depend on j and redefining it as q_j^g in an obvious way. Assumptions (ii)–(iv) could be dealt with by making the appropriate variables *functions* – and we will do this later. The single market centre assumption will simply disappear because we can handle any number of centres $i = 1, 2, 3, \ldots$ in this representation.

Both Lefeber (1958) and Stevens (1968) reformulated the von Thunen model as a linear programming model. In the notation defined above, it can be written as

$$\text{Max}_{\{Z\}}\Pi = \Sigma_{jg}(p_i^g - c_{ij}^g - v_j^g)Z_j^g \quad (7.2)$$

such that

$$q^g Z_j^g = l_j^g \quad (7.3)$$

$$\Sigma_g l_j^g = L_j \quad (7.4)$$

with the non-negativity conditions

$$Z_j^g \geqslant 0 \quad (7.5)$$

and

$$L_j^g \geqslant 0 \quad (7.6)$$

Implicitly, a single market has been assumed because there is no interaction variable and no summation over i in equation (7.2). The von Thunen model is recognisably present since the rent which landowners can collect in zone j is clearly the profit

$$r_j = p_i^g - c_{ij}^g - v_j^g \quad (7.7)$$

There is a 'free' subscript, i, on the right-hand side of the equation *because we have assumed a single market i*. However, we begin to recognise the advantage of the discrete zone system and this notation, because there is no need for the restriction to a single market. We can, of course, assume a single market, and then, essentially, von

Thunen's results can be reproduced. To generalise the model, we simply need to replace the Z_j^g term in the objective function with the interaction array, Y_{ij}^g, defined earlier, and sum over i:

$$\text{Max}_{\{Y\}}\Pi = \Sigma_{ijg}(p_i^g - c_{ij}^g - v_j^g)Y_{ij}^g \quad (7.8)$$

with, of course,

$$Z_j^g = \Sigma_i Y_{ij}^g \quad (7.9)$$

There would be a non-negativity constraint on the Y variables; the other constraints are as before.

We can, therefore, find a more general model which reproduces von Thunen. However, there is a critical further step. The generalised multi-market Lefeber–Stevens model suffers from the same defect as the transportation problem of linear programming in relation to spatial interaction models: relative to realistic situations, there are too few non-zero flows (cf. Key idea 6.7). Essentially, by adding an entropy term into the objective function, this defect can be remedied, as we have seen in constructing our archetypal models. However, we can then proceed a stage further using the methods introduced in Chapter 6: representing Y_{ij}^g through a spatial interaction model and determining Z_j^g through adding appropriate dynamic-model hypotheses. These ideas were originally developed in Wilson and Birkin (1987) and it is shown in that paper that many alternative models can be constructed within the framework that has been established. For illustrative purposes here, we focus on one of these – the one which reproduces von Thunen's model as a special case. This will illustrate the approach and show how we can generalise the classical models. However, other variants of this model may turn out better in practice. This obviously raises ongoing research issues.

Von Thunen assumes that everything which is produced is sold in the market. This suggests that we need an *attraction-constrained* spatial interaction model to reproduce this assumption, and we will also need to define a term which measures the attractiveness for farmers of a particular market centre, i for good g. Call such a term W_i^g. (Note that this is the opposite way round to the more customary production-constrained model used in the retail case where the W attractiveness term would be j-dependent rather than i-dependent.) Then an appropriate interaction model, using obvious developments of the notation would be

$$Y_{ij}^g = B_i^g W_i^g Z_j^g \exp(-\beta c_{ij}^g) \qquad (7.10)$$

where B_i^g is a balancing factor to ensure that

$$\Sigma_i Y_{ij}^g = Z_j^g \qquad (7.11)$$

so that

$$B_j^g = 1/\Sigma_i W_i^g Z_j^g \exp(-\beta^g c_{ij}^g) \qquad (7.12)$$

The next step in the argument is to embed the interaction model within a framework which handles the dynamics of $\{Z_j^g\}$. We have to assume a set of starting values for the $\{Z_j^g\}$, and then we can calculate the profit accruing to farmers in j for good g:

$$\Pi_j^g = D_j^g - C_j^g \qquad (7.13)$$

The revenue has to be estimated by summing the flows from each market centre, i, and we can deduct the costs in the process:

$$\Pi_j^g = \Sigma_i (p_i^g - c_{ij}^g - v_j^g) Y_{ij}^g \qquad (7.14)$$

We now need a mechanism for calculating ΔZ_j^g. In the spirit of von Thunen, we want to increase Z_j^g (and the land area needed to produce it) if g is the most profitable crop and we do this by taking away cultivation from less profitable crops. We can do this as follows:

$$\Delta Z_j^g = \Sigma_{h \neq g} \delta_j^{gh} \varepsilon^h q^h Z_j^h / q^g \qquad (7.15)$$

where

$$\delta_j^{gh} = 1 \text{ if } \Pi_j^g > \Pi_j^s \quad \text{for all } s \neq h \qquad (7.16a)$$
$$= -1 \text{ if } \Pi_j^g < \Pi_j^s \quad \text{for all } s \neq g \qquad (7.16b)$$
$$= 0 \text{ otherwise} \qquad (7.16c)$$

Equation (7.16a) holds if g is the most profitable crop and land is transferred to g and Z_j^g is increased; equation (7.16b) holds if there is a more profitable crop than g, and land is transferred away. The two equations are complementary. Equation (7.16c) picks up occasional cases of equality.

It can be shown in numerical experiments (Wilson and Birkin, 1987) that if this model is run with a single market centre, then the von Thunen ring pattern can be reproduced for suitable parameter values. This is shown in Figure 7.1.

However, we can now run such models for more than one market centre and a variety of patterns can be obtained, as shown in Figure 7.2. We now, therefore, have a more general model, with multiple market centres and with a spatial interaction model to represent the flows rather than a linear programming model.

It is important to emphasise, however, that a much greater variety of models can be generated. q^g could be replaced by a q_j^g and it could be used to represent varying fertility. Z_j^g could be made a function of inputs, say represented by the expenditure on production, v_j^g. Economies of scale could be incorporated in the cost function, C_j^g. We should also recall that in presenting this model we have retained (in order to show that von Thunen's model can be reproduced precisely as a special case) what was referred to in Wilson and Birkin (1987) as a hypothesis that rent is determined *passively*: it does not explicitly appear in the cost function in equation (7.13); it is assumed that the farmers' normal profits are included in the v_j^g term and that the landowner extracts the rent. However, rent could be determined actively, and this leads to an alternative model.

What this illustration and commentary show is that once an appropriate framework has been established, generalisation is straightforward in principle if not in practice!

> **KEY IDEA 7.1**
>
> A general interaction–location model of agricultural land use can be developed, which has the von Thunen model and the Lefeber–Stevens model as special cases.

Industrial location

As in the agricultural case, the first attempt to generalise the classical problem – in this case the Weber problem – was in terms of a mathematical programming model. Significantly, again, this is based on a shift to a discrete zone system or, in this case, an equivalent, a set of points which are sometimes thought of as the nodes of a network but can equally well be thought of as zone centroids. As usual, the key to generalising the problem – once

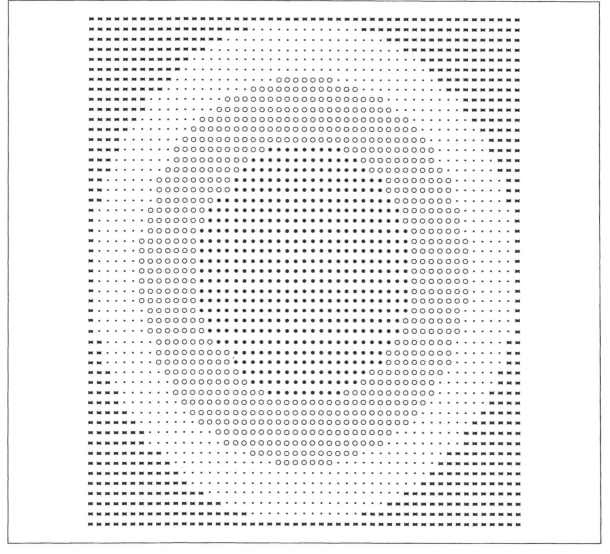

Figure 7.1 Reproduction of von Thunen's rings with an interaction model. (From Clarke and Wilson, 1985b)

the shift to a discrete space representation has been achieved – is the notation. We follow the argument from Wilson (1985a), beginning with the simplest generalised Weber problem, the so-called p-median problem (a classic in operational research), which can then be extended.

Suppose we have a set of zones or points $i = 1$, 2, 3, ... at each of which there is a set of materials, O_i, to be assigned to p factories located at possible

points $j = 1, 2, 3, ...$ (The i-set and the j-set can be the same or different depending on the problem.) Define an allocation variable λ_{ij} which is 1 if i is assigned to j and 0 otherwise. This is equivalent to defining an interaction variable

$$Y_{ij} = O_i \lambda_{ij} \tag{7.17}$$

Let x_j be 1 if there is a factory at j; 0 otherwise. Let c_{ij} be the unit cost of shipping from i to j. Then the

Figure 7.2 A von Thunen system with multiple market centres. (From Clarke and Wilson, 1985b)

optimisation problem is to find p sites which minimise the total transport cost (say, C).

$$\text{Min}_{(\lambda, x)} \, C = \Sigma_{ij} O_i \lambda_{ij} x_j c_{ij} \qquad (7.18)$$

such that

$$\Sigma_j \lambda_{ij} = 1 \qquad (7.19)$$
$$\Sigma_j x_j = p \qquad (7.20)$$
$$\lambda_{ij}, x_j = 0 \text{ or } 1 \qquad (7.21)$$

This problem is more useful in optimally locating public facilities than in industrial location modelling, though its connections with the Weber problem are clear. This is because there is no semblance of input–output relationships – clarifying what is final demand, and guaranteeing that the correct quantities are delivered to each factory. However, this is easily remedied, showing again that as soon as there is a clear statement of a problem in a good notation, it is usually possible to solve it.

Let N_{Im} be a set of *input* nodes for the supply of material m and N_C a set of *consumption* nodes with an amount X_j (of a single product) demanded at the jth of these. Let N_F be the set of possible factory nodes and let Z_j be the total amount produced in factory j. Suppose there is no constraint on supply. Let w^m be the weight of the mth material needed for a unit amount of the final product. As in the p-median problem, there are p factories to be located: $\lambda_{ij}{}^m$, $i \in N_I$, $j \in N_F$ allocates supply points to factories; μ_{jk}, $j \in N_F$, $k \in N_C$, allocates factories to final demand points. The following optimisation problem is then appropriate:

$$\text{Min}_{\{\lambda_{ij}{}^m, \mu_{kj}, x_j\}} C = \Sigma_{mi \in NIm} w^m \lambda_{ij}{}^m Z_j x_j c_{ij}{}^m$$
$$+ \Sigma_{jk \in NC} Z_j x_j \mu_{jk} c_{jk} \qquad (7.22)$$

subject to

$$\Sigma_j \lambda_{ij}{}^m = 1, \ i \in N_{Im} \qquad (7.23)$$
$$\Sigma_j \mu_{jk} = 1, \ k \in N_C \qquad (7.24)$$
$$\Sigma_j x_j = p \qquad (7.25)$$
$$\Sigma_j Z_j x_j \mu_{jk} = X_k, \ k \in N_C \qquad (7.26)$$
$$\lambda_{ij}, \mu_{jk}, x_j = 0 \text{ or } 1 \qquad (7.27)$$

Precisely as in the agricultural case, this kind of formulation suffers from being unrealistic in relation to the number of non-zero interaction flows which it offers. We can therefore proceed in what is becoming the standard way to incorporate a spatial interaction model within. However, we can now also take the argument a stage further and incorporate a full set of input–output relationships into the model and this allows us to remove the restriction of having only one final demand good.

This is a more complicated example because of the need to specify the input–output relations between industrial sectors: the process for producing a good g (as an output) will need another set of goods, $\{h\}$, as inputs. However, as in the previous example, the important starting point is system description and building up the appropriate notation. In the following, we largely follow the argument first presented by Birkin and Wilson (1986a, b), simplifying slightly for ease of presentation.

We begin by separately labelling sectors (m or n) and goods (g or h): a sector can produce several goods; and one good may be produced in more than one sector. Let $Z_j{}^{ng}$ be the amount of g

produced in sector n at j. Note that we now drop the notion of sectors which supply materials to factories – or, rather, we generalise it. Here, a materials sector is simply another production sector – and this is helpful in another sense because, of course, such sectors will also need inputs of labour, machinery and so on. There are two kinds of markets: the supply of goods to other firms as inputs to those firms and the supply of goods to final demand. As in a previous case in Chapter 6, we use set-theoretic notation to distinguish these sectors. Let N_p denote the main producing sectors; and N_{fd} the set of sectors which represent final demand. Then market sizes can be represented by $X_j{}^{ng}$ and are given by

$$X_j{}^{ng} = \Sigma_h a^{ngh} Z_j{}^{nh}, \quad n \in N_p \qquad (7.28a)$$
$$X_j{}^{ng} = F_j{}^{ng}, \quad n \in N_{fd} \qquad (7.28b)$$

where a^{ngh} is one of a set of input–output coefficients: it is the amount of good h needed to produce a unit of good g within sector n in zone j. $\{F_j{}^{ng}\}$ is a set of final demands.

As ever, the specification of the interaction term is crucial. Let it be $Y_{ij}{}^{mng}$, the flow of g from sector m in i to sector n in j. g is an output from m and an input to n. If $n \in N_p$, this will be an intermediate flow; if $n \in N_{fd}$, a flow to final demand. It is also sometimes appropriate to disaggregate further by specifying the good, in the case where n is a production sector, for which g is an input. Let $\Psi_{ij}{}^{mngh}$ be the flow of g from m in i to n in j to a process which produces h. These rather complicated array definitions capture all the relevant flows and examples are shown in Figure 7.3.

We can show more precisely what is happening in a production process for h in n at j if we think of all the inputs (g, m) from zones i ($Y_{ij}{}^{mnh}$), and the sales to sectors r in zones k for production of goods, q ($\Psi_{jk}{}^{nrhq}$). We can show the formal functional relationships as a step towards building a formal model. However, we need to make some further definitions. Let $c_{ij}{}^g$ be the unit cost of shipping g from i to j. Then we need a measure of the attractiveness of sector m in i as a supplier of g to n in h. Call this $U_{ij}{}^{mng}$ (assuming it to be independent of h). Formally, we can write

$$U_{ij}{}^{mng} = U_{ij}{}^{mng}(Z_i{}^{mg}, p_i{}^{mg}, c_{ij}{}^g, \dots\dots) \qquad (7.29)$$

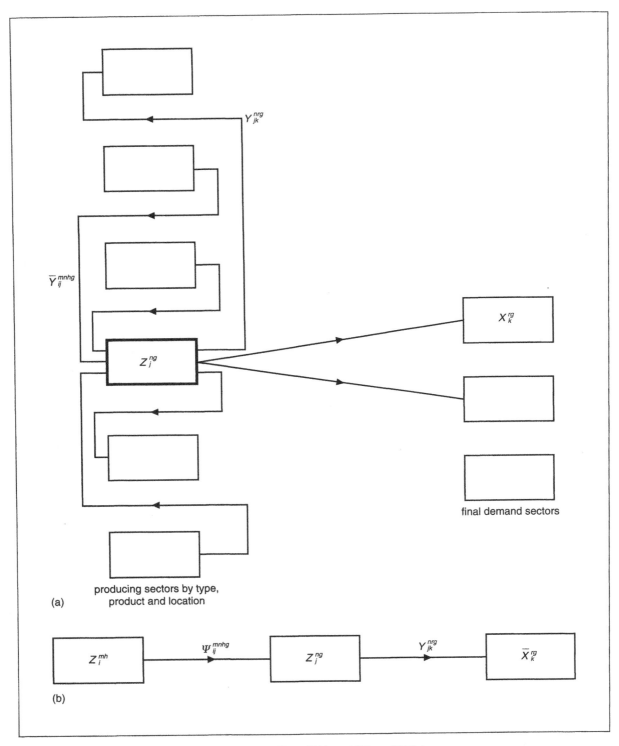

Y_{jk}^{nrg}

$\overline{Y}_{ij}^{mnhg}$

Z_j^{ng}

X_k^{rg}

final demand sectors

producing sectors by type,
product and location

(a)

(b)

Z_i^{mh} → Ψ_{ij}^{mnhg} → Z_j^{ng} → Y_{jk}^{nrg} → $\overline{X}_k^{rg}$

Figure 7.3 Key arrays in urban input–output processes. (From Birkin and Wilson, 1986a)

showing the attractiveness to be a function of the total amount produced, the price, the transport cost and so on. (This form of specifying formal relationships is a useful intermediate step in model-building.)

The input flows to (n, h, j) can then be written, still in the formal mode, as

$$Y_{ij}^{mnh} = Y_{ij}^{mnh}(U_{ij}^{mnh}, \Sigma_g a^{ngh} Z_j^{ng}, c_{ij}^h) \qquad (7.30)$$

(where $\Sigma_g a^{ngh} Z_j^{ng}$ is the total amount of h needed as inputs to all production processes in j). We can further disaggregate and focus on a particular process g in n in j by removing the summation sign in equation (7.30) and writing it in terms of the Ψ-array:

$$\Psi_{ij}^{mnh} = \Psi_{ij}^{mnh}(U_{ij}^{mnh}, a^{ngh} Z_j^{ng}, c_{ij}^h) \qquad (7.31)$$

The outputs, on a similar basis (i.e. g to sector r in zone k) are

$$Y_{jk}^{nrg} = Y_{jk}^{nrg}(U_{jk}^{nrg}, X_k^{rg}, c_{jk}^g) \qquad (7.32)$$

where X_k^{rg} will be an intermediate total or final demand as appropriate (as specified in equations (7.28)).

Suppose $\{p_j^{ng}\}$ is a set of prices for good g produced in sector n in zone j. Then, assuming that the spatial interaction models can be fully specified, we can calculate

$$D_j^{ng} = \Sigma_{kr} p_k^{rg} Y_{jk}^{nrg} \qquad (7.33)$$

as the revenue attracted to the (g, n) process at j. The cost of production can be taken as

$$C_j^{ng} = f_j^{ng} + v_j^{ng} Z_j^{ng} + \Sigma_{imh}(p_i^{mh} + c_{ij}^h)\Psi_{ij}^{mnhg} \qquad (7.34)$$

where f_j^{ng} represents the fixed costs, v_j^{ng} the variable costs, and the final term is the cost of the inputs. We can then, as usual, find the equilibrium distribution $\{Z_j^{ng}\}$, by solving

$$D_j^{ng} = C_j^{ng} \qquad (7.35)$$

In the usual way, we can also write down equations for the dynamics:

$$\Delta Z_j^{ng} = \varepsilon^{ng}(D_j^{ng} - C_j^{ng})Z_j^{ng} \qquad (7.36)$$

It is always an interesting test of a more general model to reproduce any original special cases. If we take three N_p sectors, (1, 2, 3) and one final demand (N_{fd}) sector (4), and specify the input–output coefficients so that 1 and 2 supply

inputs to 3, and 3 supplies final demand to 4 (each of 1, 2 and 3 producing a single good) then we reproduce Weber's problem for this discrete zone case. Figure 7.4 then shows how the general model does indeed precisely reproduce Weber's results, the different plots showing the location of sector 3 coinciding with the fixed locations of 1, 2 and 4 in appropriate circumstances; or being located in the triangle. It is similarly possible to reproduce versions of Palander's or Hotelling's models. These results are presented in Birkin and Wilson (1986b). Clearly, however, it is also possible to be much more general. Figure 7.5, for example, shows some results of model runs for five linked sectors with different assumptions about population density (and hence the spatial distribution of final demand).

As in the case of the agriculture model, it is also possible to generalise further on the basis of the framework that has been developed. For example, Z_j^{ng} could be made a production function and C_j^{ng} a more general cost function. Prices could be determined by some endogenous mechanism within the model or by a dynamic adjustment equation.

KEY IDEA 7.2

A general interaction–location model of industrial location can be developed which incorporates the key hypotheses of diverse earlier models – ranging from Weber via Hoover and Palander to input–output modelling. It provides a rich framework for further development.

Residential location and housing

Residences are land-consuming and so residential land use competes with agricultural land use – the other large consumer. In one sense, therefore, housing could be added to the agricultural model as another 'sector'. It is not surprising, therefore, that Alonso in effect did this in using his *bid rent*

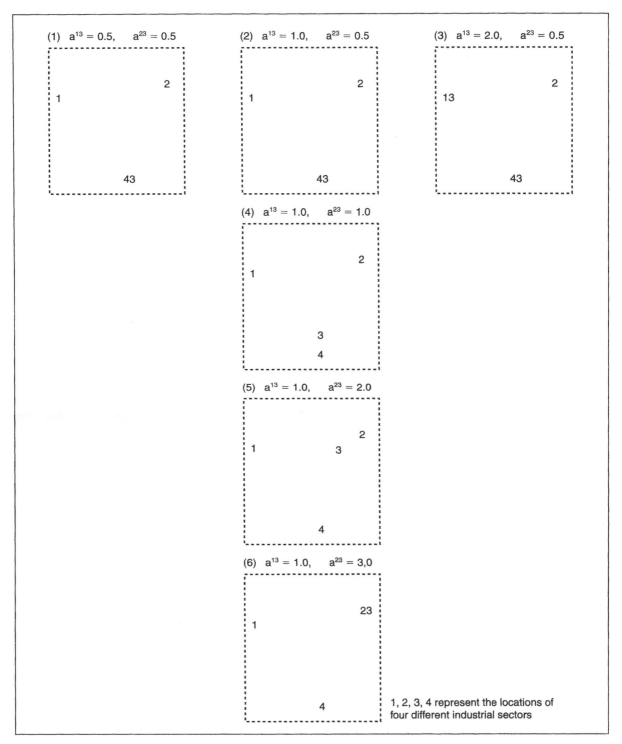

Figure 7.4 Weber's results from an interaction model. (From Clarke and Wilson, 1985b)

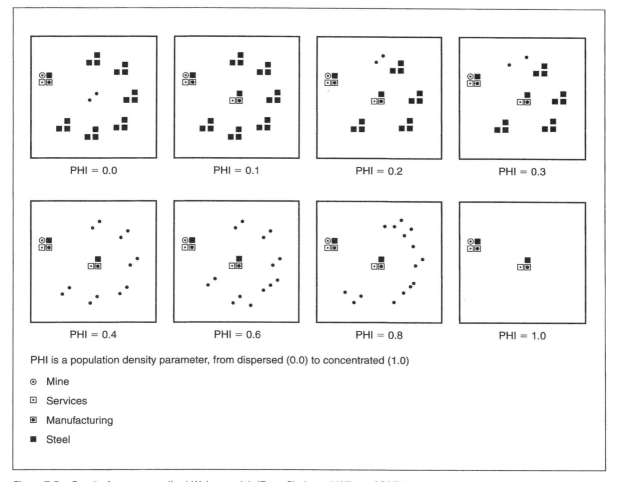

Figure 7.5 Results from a generalised Weber model. (From Clarke and Wilson, 1985b)

concept. However, we are also, of course, interested in structures within residential areas (the equivalent of different crops?) and we saw in Chapter 5 that there had been empirical approaches to this question by Burgess (emphasising a ring structure), Hoyt (a sectoral structure) and Harris and Ullman (a multi-nucleated structure). This is clearly a situation that cries out for an integrated model which can reproduce any of these patterns – or combinations of them – exactly as we have managed to do in the cases of agricultural and industrial location. The argument proceeds in a similar way and it is interesting in relation to the history of the development of ideas that yet again, the first

approach using a new and more general notation involved a mathematical programming model. For completeness and interest we begin here too, and then generalise this through the use of a spatial interaction model.

Alonso applied von Thunen's concept of bid rent in a residential location situation: he showed how bid rents could be used to represent preferences for housing for different kinds of people, and that a market could be cleared and that a land-price surface could be determined. However, he was operating with a continual space representation and it was difficult to make the model operational. Essentially because they switched to a discrete zone representation and considered people in groups, Herbert and Stevens (1960) were able to construct an operational model. This can be presented as follows.

Let T_i^{kw} be the number of type w people (or, more precisely in this case, households) in type k houses in zone i; let b^{kw} be the bid rent for type w people for type k housing; let p_i^{kw} be the price paid *exclusive of site cost*. Let q^{kw} be the average amount of land used by a type w household for a type k house; suppose there are P^w type w households and the total land area in zone i is L_i. Then, their model is

$$\text{Max}_{\{T\}} \; B = \Sigma_{ikw}(b^{kw} - p_i^{kw})T_i^{kw} \tag{7.37}$$

subject to

$$\Sigma_{kw} \, T_i^{kw}q^{kw} < L_i \tag{7.38}$$

and

$$\Sigma_{ik} \, T_i^{kw} = P^w \tag{7.39}$$

There are clever ideas embedded in this model – notably the representation of bid rents through b^{kw}, and the fact that the dual variables represent rents. However, it suffers from the same defect as the transportation problem of linear programming I used as a journey to work model: it is too perfect; it does not represent real-life dispersion. This can be rectified through procedures which are now familiar – through the introduction of a spatial interaction variable and an associated model. (We should also note that the dispersion could be added directly without the interaction element.)

For simplicity, we assume one worker per household (which can easily be relaxed later) and, more importantly, that we should give the journey to work prominence in establishing residential location. Then, extending the previous notation: let T_{ij}^{kw} be the number of households whose principal worker works in zone j for a type w job (assuming that type of job is characterised for these purposes by income) and who live in a type k house in zone i. Let H_i^k be the number of type k houses in zone i; E_j^w the number of type w jobs in zone j; and W_i^{kw} the attractiveness of type k houses in zone i for type w people. Then, with B_j^w as the usual balancing factor, the basic spatial interaction model which assigns workers to available housing is:

$$T_{ij}^{kw} = B_j^w W_i^k E_j^w \exp(-\beta^w c_{ij}) \tag{7.40}$$

where

$$B_j^w = 1/\Sigma_{ik}W_i^k \exp(-\beta^w c_{ij}) \tag{7.41}$$

With a suitable form of the model, we could then derive the Herbert–Stevens model by letting $\beta^w \to \infty$.

Let p_i^k be the (annualised say) price of a type k house in zone i and let s^w be the average amount spent by a type-w household on housing. Then

$$C_i^k = p_i^k H_i^k \tag{7.42}$$

is the amount spent on type k housing in i and

$$D_i^k = \Sigma_j{}^w T_{ij}^{kw}s^w \tag{7.43}$$

is the average amount of cash available. The usual mechanism would then suggest

$$\Delta H_i^k = \varepsilon^w(D_i^k - C_i^k) \tag{7.44}$$

for suitable parameters ε^w and

$$D_i^k = C_i^k \tag{7.45}$$

is the usual equilibrium condition.

The real situation is very much more complicated than this. The attractiveness term can be specified as a series of factors, for example. In Clarke and Wilson (1983), five such factors are used, representing (i) housing supply, H_i^k; (ii) access to services; (iii) an attractiveness factor in relation to other social groups; (iv) a repulsion factor in relation to other social groups; and (v) a term relating to house prices. The detail need not concern us here.

A variety of equilibrium patterns can be obtained with this model, as shown in Clarke and Wilson (1983). It is easy to see how rich combinations of ring, sector and multiple-centrality can be obtained; and so the model can reproduce and then generalise the Burgess–Hoyt–Harris and Ullman theories.

KEY IDEA 7.3

A general interaction–location model of residential location can be developed which can reproduce within its predictions mixes of ring, sector and multiple-nuclei structures.

Services

Since we have used the example of the service-sector model as the main example to illustrate the development of the argument in Chapter 6, we can be relatively brief here. However, this argument can be considered as a first-stage replacement for central place theory; the second stage will be supplied when we review briefly the tasks involved in assembling a comprehensive model later in the chapter.

We have already seen that we can generate a variety of equilibrium patterns. The next key step is to indicate how we can disaggregate by type of good, say g. If we do this in the most straightforward way, the equations become

$$S_{ij}^g = A_i^g e_i^g P_i W_j^{g^\alpha} \exp(-\beta c_{ij}) \qquad (7.46)$$

where

$$A_i^g = 1/\Sigma_j W_j^{g^\alpha} \exp(-\beta c_{ij}) \qquad (7.47)$$

to ensure that

$$\Sigma_j S_{ij}^g = e_i^g P_i \qquad (7.48)$$

This is simply a series of models of the type we have explored extensively above – one for each g. The sets of parameters (α^g, β^g) can be used to reproduce the essential characteristics of CPT structures: high-order goods in the hierarchy will have higher α and lower β; and vice versa. However, there is something missing: there is no advantage in the formulation to centres which supply a range of goods. This can easily be represented by using the same kind of technique as in the residential location model: by modifying the attractiveness function so that it is made up of a number of factors one of these can then represent agglomeration weighted by a parameter that allows us to represent the importance of these effects. The parameter can be determined empirically.

KEY IDEA 7.4

The interaction–location model of services can be developed to apply to a wide variety of circumstances. Its outcomes will reflect the hierarchical structures of central place theory without giving any preference to particular geometries – such as hexagonal structures.

Transport

Before proceeding to the issue of integrating submodels, it is useful to add a note on the transport model. We have already seen in our historical review of modelling that it was the context of transport engineering which generated the doubly constrained spatial interaction model; and that urban and regional analysts have learned to use this, and other members of the family of models, as part of the corpus of their own theory. However, if the prime focus is transport flows, then the basic model can be extended – by linking it to other important transport submodels and by disaggregation.

It is useful to think of the transport model as made up of four submodels: trip generation, trip distribution, modal split and assignment. In the basic model (neglecting modal split initially), the trip generation submodel estimates the origin and destination totals, $\{O_i\}$ and $\{D_j\}$ – for example as functions of land use and associated activities. The trip distribution model estimates the basic spatial interaction array, $\{T_{ij}\}$. The assignment model loads the trips on to the transport network and it is at this point that we can compute better estimates of travel time (for example, in relation to congestion) and hence the array $\{c_{ij}\}$. This implies that the model has to be run iteratively within itself: estimate $\{O_i\}$ and $\{D_j\}$; then $\{T_{ij}\}$ as functions of $\{c_{ij}\}$; then re-estimate $\{c_{ij}\}$ in relation to network loadings and repeat the sequence.

Clearly, in order to introduce transport mode, k, then it is necessary to disaggregate. It is also necessary to have a person type index, n, at least to distinguish car owners from non-car owners, since obviously the latter group will not have the car mode available to them. The main array of variables to be predicted is then $\{T_{ij}^{kn}\}$. The trip generation model predicts $\{O_i^n\}$ and $\{D_j\}$, trip origins being distinguished by n, trip destinations not; and we have to introduce a set of modal cost, $\{c_{ij}^k\}$ by disaggregating $\{c_{ij}\}$. The distribution and modal split submodels can be run jointly or consecutively. The assignment model loads trips on to modal networks, and then adjusts $\{c_{ij}^k\}$ to take account of congestion. The main disaggregated interaction model becomes

$$T_{ij}^{kn} = A_i^n B_j O_i^n D_j \exp(-\beta^n c_{ij}^k) \qquad (7.49)$$

and, of course, this determines the proportion travelling by mode k:

$$M_{ij}^{kn} = \exp(-\beta^n c_{ij}^k)/\Sigma_{k\varepsilon\gamma(n)} \exp(-\beta^n c_{ij}^k) \quad (7.50)$$

where the range of summation, $k\varepsilon\gamma(n)$, is over modes available to people of type n. That is, $\gamma(n)$ is defined as the set of modes available to people of type n.

One more element of real-world complexity has to be added in practice: trips need to be distinguished by purpose, p; so that the main array becomes $\{T_{ij}^{knp}\}$. It is then useful to think of the interaction model as being doubly constrained for some purposes, such as the journey to work; and singly constrained for others, such as shopping. This can then provide a more effective link to the other submodels.

KEY IDEA 7.5

The basic interaction model can be developed as a disaggregated transport model. Its equivalent locational component would involve modelling the development of infrastructure such as roads. (See Wilson, 1974, chapter 9, for an example.)

Urban and regional structure: integrated comprehensive models

It will already be clear that we could in principle connect together the submodels we have discussed to build a comprehensive model; and we noted that this would form the second leg of the task of reconstituting central place theory. Indeed, this has been a long-standing research task in the urban modelling field, with the Lowry (1964) model as the precursor of a whole generation of comprehensive models. The structure of Lowry's original model is shown in Figure 7.6 and this demonstrates some of the features that have been in most models since. He assumed a 'basic' (essentially industrial) employment sector whose spatial distribution was given. Its workers had to be allocated to residences; these residents then demanded services and thus created more

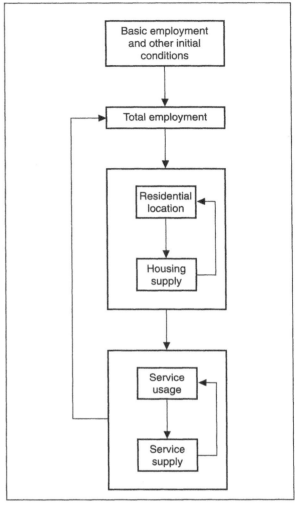

Figure 7.6 The structure of the Lowry model. (From Wilson and Bennett, 1985)

employment. This in turn was reallocated to more residences; and so the iteration continued. In principle, it is now possible to add in explicitly an agricultural submodel (particularly in relation to land use at the rural–urban fringe) and an industrial model.

We can think of the core situation as one of *interacting fields*: fields emanate from employment centres of workers seeking houses; the concentrations of residents demand services and hence create more employment. These interactions are likely to change the patterns of evolution. The kinds of results which can be obtained from an extended Lowry model are presented in Birkin *et al.* (1984). In the submodels presented in earlier sections, in each case we have taken important arrays as given for that submodel: the residential distribution is given for the retail model and the employment distribution is given for the residential location model. What we now see is that the models (with others) should be run on an interactive basis, say year by year.

There have been many attempts to specify a comprehensive model over the 35 years since Lowry's paper was published. It takes us beyond the scope of this book to present any of them in detail. We briefly review the progress that has been made in the first section of Chapter 8.

KEY IDEA 7.6

The component models presented in earlier sections can be seen as components of an integrated model. It is possible in principle to write the whole model system in a general notation so that the sectors which have been identified – agricultural, industrial, residential and service can each be seen as instances of a more general 'activity sector'.

Concluding comments

The ideas presented in this chapter are exciting in a number of respects. The theoretical tools offered by spatial interaction modelling and dynamical locational analysis 'solve' many of the problems faced by the classical theorists. Restrictions can be removed. The hypotheses of the classical theorists can be incorporated in the frameworks and their results reproduced in both restricted and extended forms. Alternative hypotheses can be explored easily once the principles have been mastered. There is now the possibility of a vastly extended programme of empirical testing and this would inevitably generate new theoretical developments.

8

The research agenda

The starting point: experience to date

We have seen that there is a set of models which provide the basis for urban and regional analysis and which herald major advances in our understanding of complex spatial systems. The models can function separately, or they can be linked into a comprehensive model. The emphasis so far has been on the theoretical development of these models. In this section, we offer a brief account of the experience of using the models – mainly through reference to recent literature – as a preliminary for a review of the ongoing research agenda. It will be argued that there are tremendous opportunities for the future, and these are pursued in the rest of the chapter.

Experiences to date are captured in a number of recent reviews: for example, see Batty (1994), Berechman *et al.* (1996), Bertuglia *et al.* (1987), Harris (1994), Klosterman (1994), Wegener (1994) and Wilson (1998). Klosterman (1994) edited a special issue of the *Journal of the American Institute of Planners* that reviewed Lee's (1973) paper which had been published as a 'requiem for large-scale models'. Fortunately, the reviewers are able to report substantial progress rather than any agreement with Lee's original article!

The models have been most extensively used in the transport and the retail fields. In the transport case, their use has become almost routine. The models have been refined, often in response to practical requirements to represent transport systems at different levels of detail and complexity. The four-stage submodel described in Chapter 6 is now often divided into different combinations of submodels to represent specific approximations in different applications' contexts. For example, if a short-run forecast of the effects of a network change is needed, the model may have a fixed trip matrix, say from a survey, and the focus would be on the assignment model. However, there remain research issues: for example, the task of estimating generated traffic following new investment.

In the retail case, most of the applications have been in the private sector. This has been facilitated by the growth of companies such as GMAP Ltd. (part of which is now trading as Polk Ltd.) whose work is described in Birkin *et al.* (1996). This work has had the advantage of making some rich data sets available for model development and testing – and, in effect, has been the basis for a substantial research programme on model testing. It should also be noted that 'retail' should be given a broad interpretation in this context: GMAP's work, for example, includes applications in the fields of health, the motor industry, water resources and financial services as well as retailing more narrowly defined in terms of shopping.

> **KEY IDEA 8.1**
>
> GMAP Ltd. has demonstrated the power of models for commercial application.

There do not seem to have been separate developments in areas such as the modelling of agriculture, industry or residential location except in the context of the building of comprehensive models (and even in that context, with relatively little focus on agriculture and industry). There has been almost no serious developmental work on the dynamic modelling ideas presented in this book.

Most of the applications have focused on the relatively short run, and forecasting is typically on a *comparative static* basis. There are, of course, fundamental difficulties in this case, and we return to these later in the chapter.

Wegener (1994) estimates that there are about 20 research groups around the world who have continued to develop large-scale comprehensive urban models, mainly in universities, but with many working in applied contexts. There have been many applications in the United States, mainly led by Putman (1983, 1991) using an extended Lowry model with entropy-maximising submodels. Interestingly, these applications have been driven in part by the requirements of air quality legislation which require pollution impact studies to be carried out in the context of major transport proposals, and this in turn demands consideration of land-use–transport interaction, and hence the deployment of comprehensive models. There have also been a significant range of applications in Europe, notably by Echenique and his colleagues (e.g. Echenique *et al.*, 1990); and these activities have been extended to South America by de la Barra (1989, for instance). A number of European academics have continued to develop comprehensive models and many have collaborated (together with Japanese and American colleagues) in the ISGLUTI Project, described by Webster *et al.* (1988) and Paulley and Webster (1991). There are further details in papers by Mackett (1980, 1990, 1993), Wegener (1986a, b) and Nakamura *et al.* (1983). The range of dates on these papers show the time commitment and stamina that has proved necessary for this work. There has been another long-running international collaboration centred on Bertuglia's group at the Polytechnico Torino. This research has had a mainly theoretical drive, though models have been calibrated for Italian cities. Their work is collected in a series of books (Bertuglia *et al.*, 1987, 1990, 1994).

KEY IDEA 8.2

The importance of the comprehensive model-building exercise continues to be recognised.

Notwithstanding the volume of work indicated, it can be argued that the hopes of the modelling communities in the 1960s have not been fulfilled – except possibly in transport and, to a lesser extent, in retailing (cf. Wilson, 1985c). This is in part due to changes in fashion: from about the mid-1970s, interest in public planning declined, a process well documented for the modelling context by Batty (1989, 1994). However, fashions change, model implementation costs decrease as computing costs decrease, and the efficacy of modelling in areas where it has been applied may well increase levels of interest in other areas. There may be grounds again, therefore, for optimism. This will be a component of our analysis of the ongoing research agenda to which we now turn.

Research opportunities

INTRODUCTION

In the rest of this chapter, we seek ways of articulating the research agenda for the future. We first focus on the short run, i.e. opportunities that are available now. It should be borne in mind at the outset that these can range in scale from undergraduate projects through PhD theses and Research Council projects to a new kind of 'big science'. In the rest of this introductory section, we review what can be achieved from the application of the STM principles for model-building articulated in Chapters 2, 3 and 4. In the next two sections, we draw together the opportunities arising from methodological developments, and then review the question of the application of urban and regional models in planning and problem-solving. These two sections then provide the basis for a section that focuses on analysis-based model-building. We review the range of application to a variety of systems, giving particular emphasis to systems and problems where modelling is seriously under-developed. We draw as appropriate on earlier sections for new ideas to apply in particular contexts. In the following section, we examine some research challenges for the longer run, taking in turn the conceptual, the mathematical and the computational. The research agenda raises substantial questions about how it can be implemented, and these are pursued in the last section.

APPLICATIONS OF THE PRINCIPLES

In Chapters 2, 3 and 4, we set out some principles for model-building for urban and regional analysis, consideration of which, it can be argued, will lead to productive research projects. The core of these principles is the STM framework: *system* articulation, *theory* development and the assembly of appropriate *methods*. There are almost endless possibilities for generating new perspectives. These have been illustrated in Chapters 5 and 7 which have shown how the classical models of geographical theory can be rewritten. This turns on redefining the systems of interest and then seeking, and finding available, more powerful methods for model-building. This argument, however, was presented in general terms. Research projects could be defined that develop particular examples of these and which are perhaps particularly appropriate for academics and students.

> **KEY IDEA 8.3**
>
> Powerful model-building principles are now available which not only allow the classical models to be rewritten but provide opportunities for a multitude of research projects.

It has already been noted that very little work has been done on charting out examples of change using dynamic models. An early example of change through a critical point was the shift from corner shop retailing to a supermarket base (Wilson and Oulton, 1983). There are many other examples of such change in spatial organisation, particularly in different kinds of retailing (e.g. the superstore) and in the public services (e.g. the notion of hub and spoke systems for hospitals, with a superhospital, say a major teaching hospital, more closely connected to district hospitals). Trends could be detected through relatively easily acquired data and suitable model systems defined. We return to this example below.

> **KEY IDEA 8.4**
>
> It is important to look for key structural changes and to seek to understand these through dynamic models. This understanding will help to predict possibilities of future structural change.

An interesting specific kind of limited research project – perhaps suitable for student projects – would be to seek to measure the *complexity* of various systems of interest using the methods indicated in the section on the measurement of complexity in Chapter 4. By applying Ashby's law of requisite variety (given appropriate assumptions about which variables might be controllable), it would then be possible to understand the plannability of different kinds of system.

> **KEY IDEA 8.5**
>
> Complexity and plannability combine to produce fruitful research topics.

When interesting systems (and problems) have been defined, attention turns to theory and method. At this point, the key implication of Chapter 4 is that it is critically important to be ruthlessly multidisciplinary. Elements of theory and method for any particular system in urban and regional analysis are likely to come from several disciplines.

These tasks can be accomplished, at least in a preliminary way, before any attempt is made to build a model on a computer. This then reveals another kind of research project – again perhaps suitable for student dissertations at undergraduate or masters levels: to use the STM model framework for achieving a *qualitative* analysis of the system.

> **KEY IDEA 8.6**
>
> Model systems provide an excellent framework for qualitative analysis.

In summary, what we are seeking to achieve is analytical understanding and a demonstration of effectiveness of the model through its reproductive power: testing the outputs of the model against real data. And, of course, we are also interested in the future predictive power of the model, especially for the short run which is often itself valuable. The explorations involved in these processes will be accomplished through human-machine interaction, and for this purpose, model outputs can be integrated with geographical information systems.

Opportunities from methodological developments

INTRODUCTION

It is worthwhile to examine areas of methodological development which either form significant parts of the research agenda in their own right or will provide part of the backcloth to substantive systems and planning research.

The model systems described in this book, in computer programming terms, have an obvious connection to the FORTRAN programming language and the methods used have typically reflected this. A combination of new programming possibilities and new conceptual ideas are likely to revolutionise this perspective in the future and it is against this background that we now review three topics. We discuss in turn: alternative ways of representing systems, 'auto' modelling and intelligent GIS and human–computer interaction.

REPRESENTATIONS: MICROSIMULATION AND OBJECT ORIENTATION

Urban and regional modelling systems are obviously concerned with large algebraic arrays which have to be represented in the computer. If these arrays, as is often the case, are largely empty, then it is tempting to seek an alternative computer representation. In this case, the alternative representation also leads to some important new modelling ideas. The notion to be considered is that of *microsimulation*. Essentially, this involves listing a hypothetical population and, against each member of this population, its characteristics, which can include, say, residential location,

workplace location and so on. It is immediately clear that a full and *dense* representation of the system that is being modelled can be achieved. The models are then used, in effect, to calculate the probabilities of particular characteristics being associated with a particular individual, and some random number device can be used to choose the actual characteristics in such a way that the distributions of these in the simulated population match the real or modelled distributions.

Microsimulation has a long history, its origins usually being attributed to Orcutt (1957; see Orcutt *et al.*, 1968). The origins of its use in model systems based on interaction–location concepts are in the paper by Wilson and Pownall (1976) and the method has been extensively developed in Leeds since (e.g. Clarke *et al.*, 1981; Duley and Rees, 1991). The methods are now being applied widely in disciplines associated with urban and regional analysis. See, for example, Gilbert (1995) for an example in sociology, Hancock and Sutherland (1992) and Harding (1990) for examples in economics and social administration, and Kain and Apgar (1985) and Kain (1987) for examples in housing and residential location.

Will there be a further shift that is computer programming driven? The rapidly developing literature on object-oriented programming using languages such as C++ suggest that this might be the case. Objects in these languages are essentially subsystems and there is clearly a degree of structure in the ideas which take the programmes beyond those of relational databases and list processing. These ideas extend into OLAP systems: on-line analytical processing, building on the notion of multidimensional information systems and they seem to fit well with the requirements of large-scale model-building (e.g. Thomsen, 1997). A lead in geography has been provided by Leung (1997). A major research challenge is to integrate his object-oriented expert-systems approach with the analytical capability provided by the models in this book.

It is appropriate to conclude this subsection with a general comment on simulation. We have discussed briefly the microsimulation method, and we have indicated how the development of new computer languages is likely to facilitate simulation in the future. What is clear is that in many disciplines, and urban and regional modelling will reflect this, the only way to 'solve' analytically intractable mathematics is to use the power of supercomputers to simulate systems. This is itself a powerful pointer to future research methodologies. However, it also raises an issue that is a more general analogue of the research problem discussed above in relation to large numbers of performance indicators. How can we intelligently scan simulation outputs? How can we 'know' what we are looking for?

AUTO MODELLING, NEURAL COMPUTING AND NEW REPRESENTATIONS

Modellers have long been concerned with 'automatic' calibration processes (e.g. Batty, 1976). It is now possible to use the ideas of neural networks as deployed in neural computing to extend these notions, not simply for calibration, but to find the 'optimal' model which fits a set of data. This can be seen as a sophisticated kind of statistical analysis rather than mathematical modelling since, in principle, we still want to derive our model from a good theory! However, the *appearance* of models generated in this way may generate useful ideas (see Openshaw, 1988, 1992, 1993; Leung, 1997). What is clear is that this kind of research relies on the availability of super-computing power.

However, there is a possibly significant further research area here. Authors such as Deco and Obradovic (1996) articulate the concepts of neural computing in a general way – and in their particular case, helpfully build on an information-theoretic (entropy-maximising) framework. If we couple their formal developments with the more intuitive ideas presented in Holland's recent work (Holland, 1995, 1998), then we can see that, in principle, the family of spatial interaction and location models presented in Chapter 6 could formally be represented as neural models, and this would provide a more explicit framework for comparing these mathematical models with the quasi-statistical models generated by Openshaw.

To this can be added a further research task. The learning adjustment mechanisms in a neural network model are reminiscent of the $\varepsilon(D_j - C_j)$ adjustment mechanisms in dynamic location models. This suggests that it should be possible to interpret the adjustment mechanism (through the hidden 'neurones') in a neural network representation of a location-interaction model as the agents 'learning' to respond to the market situations in which they find themselves. So the formal representation may lend itself to helpful and interesting interpretations. Indeed, through the modification of the learning mechanisms, it may lead to new models (or to the more effective interpretation of Openshaw's statistical models).

The neural network approach can be considered as an alternative representation. This raises the wider prospect of what can be achieved by alternative computer *languages*. Hillis (1987, 1999), for example, discusses LISP and cmLISP in the context of parallel processing. Since parallel processing is concerned with large numbers of small memory processors interacting, and since cities, with large numbers of people and organisations are in some ways structurally similar, this may be a valuable route to explore.

IGIS, SIMULATION AND HUMAN–COMPUTER INTERACTION

The development of geographical information systems has a long tradition (for reviews, see Coppock and Rhind, 1991; Maguire *et al.*, 1991). They have increasingly been integrated with modelling systems (for reviews of these developments, see Fischer and Nijkamp, 1993; Longley and Batty, 1996). And, of course, it has been argued earlier that Birkin *et al.* (1996) is essentially about the development of *intelligent* GIS. There are two underlying general ideas here. First, that good visualisation is important and should not be underestimated; and secondly, that human–computer interaction is likely to be the main route to future progress. We saw in the previous subsection that simulation is likely to be very important in future research and that an associated research problem was the scanning of simulation outputs. Until that problem is solved – and probably even then – human–computer interaction with the intelligent scanning of intelligent GIS presentations of model outputs is likely to be the productive route forward.

> ### KEY IDEA 8.9
>
> GISs have added enormously to our capacity for visualisation (and hence interpretation) of model outputs. By linking them with model systems, they can be made intelligent. We can then combine computer power and model-based analysis with human intelligence.

Applications of models in planning and problem-solving

INTRODUCTION

Applications are typically concerned with aspects of policy, planning or problem-solving and we saw in Chapter 2 that there are three kinds of activity associated with these processes:

- policy
- invention
- analysis

The main content of the book is concerned with analysis. To explore the research problems associated with application, we can assume that we have a model-based analytical capability so that we can focus on the other two. The main purpose of the analytical capability is to provide forecasting power, and we review the research aspects of this in a later section. First, however, we focus on performance indicators as a powerful concept which allows us to define what we mean by policy. We conclude with a review of model-based research issues associated with design.

PERFORMANCE INDICATORS

For any organisation – a business or a public authority – to focus on *policy*, it needs to articulate its *objectives*. In so far as these can be expressed quantitatively, they can be represented as what we have been calling *performance indicators*. We have seen in Chapter 6 and in this chapter how examples of performance indicators can be defined. We have seen in Chapter 4 how they can be connected, for instance, to the underlying principles of cost–benefit analysis. What is perhaps remarkable, is that there is no systematic concern in most organisations with the definition of appropriate performance indicators. From time to time, there seems to be an interest in the development of management information systems (MIS) – even executive information systems (EIS) – but this is rarely sustained. There are other areas – for example in the development of the *Treasury Manual* on project appraisal – where good work has been done in providing a framework for project evaluation (which is in effect the articulation of performance indicators), but in these cases, the analytical base is often very poor. There are therefore significant research opportunities (easy pickings indeed!) in simply systematically articulating the sets of performance indicators which could represent outcomes in a variety of organisations and systems.

If this were to happen, there would almost certainly be a renewal of interest in areas such as cost–benefit analysis, which can provide much of the underlying experience, and where much experience has already been gained in tackling the difficult issues of measurement which often arise.

What seems to have happened in recent years is that 'cost–benefit analysis' has passed into common usage but in a very watered-down form. There is a temptation to revert to subjective scoring systems for difficult-to-measure indicators – and, of course, outcomes can then easily be manipulated by those defining these systems.

There is one different kind of research problem associated with performance indicators to which it is worth drawing attention. It will be clear from the health and education examples above that model-based systems can generate large numbers of indicators. For example, we have arrays with indices such as (i, m, n) or (j, m, n), where i ranges over residential zones and j over hospitals or schools; m over indices such as social class; and n over types of case or provision. These arrays provide batteries of indicators which might contain thousands or even millions of indicators. It is this situation which tempts the designers of executive information systems ('Managers can only cope with a small number of numbers ...') to attempt to encapsulate all the policy issues of a system into a small number of indicators. In the National Health Service in the UK, for example, there is a focus on waiting lists – at times almost to the exclusion of all other indicators. The research problem then is this: how can we define an *intelligent scrutiny system* for performance indicators which both recognises real system complexity but also generates fewer numbers? The key to this must be to have ways of establishing normal (or expected) bounds for performance indicators. A scrutiny system could then simply report indicators that were out of bounds. The more intelligent it was, the more it would be capable of adding possible remedial actions to its reports – but that takes us into the subject matter of the next section.

> ### KEY IDEA 8.10
>
> Much could be gained from the development of an intelligent scrutiny system for batteries of performance indicators.

FORECASTING

The distinctive component of model-based planning is the capacity for 'what if' forecasting. In any model, this capacity turns on the ability to forecast any exogenous variables and, of course, on the model's ability to predict. We distinguished earlier between forecasting capability and reproductive capability. If reproductive capability is good, then at least short-term 'what if' forecasting should be reasonable – and for many purposes, this is all that is needed (e.g. for a redeployment of hospital services which could be accomplished in the short run.) We also saw from our earlier analysis that long-run forecasting, because of the nature of the dynamics of the system, is virtually always going to be a problem. We did see, however, that it might be possible to predict *types* of development rather than the specifics, and that this could be valuable.

> ### KEY IDEA 8.11
>
> In complex dynamic situations, forecasting in detail is intrinsically impossible. However, short-run forecasting is what is needed for many planning purposes; and, for the long run, it may still be possible to forecast types of development.

THE COMBINATORIAL PROBLEM OF DESIGN

To solve problems or to make plans, it is necessary to decide which variables can be 'controlled'. The design task is then to solve problems, or to make good plans, by *inventing* appropriate settings of the controllable variables. Traditionally, in urban planning, these have been the land-use control or zoning variables (in effect the traditional town planner's 'map') and the transport networks. In an area like retailing, it is the specification of the branch network. This, of course, is a very formal way of putting the design task. In many cases, a traditional designer's or inventor's flair will be needed to find solutions. The model-based analysis can then be used in a 'what if' mode to test out these solutions or plans. With good sets of performance indicators, and possibly a means of combining them into a single measure (as in cost–benefit analysis for instance) then it may be that the best plan among alternatives can be chosen.

This general description defines such a wide set of problems that there are clearly many research opportunities. As in the case of the model-building process itself, many of the research questions will define themselves through practice: work on a model-based approach to a big issue and there will be a design element to the research. It is also appropriate to explore whether there are more formal design methods which can be brought to bear (cf. Alexander, 1964 – a valuable book which is still in print). It is also interesting to see whether there are any formal mathematical approaches – to embed the model *and* the design problem within an optimising framework. This process was systematically explored in Wilson *et al.* (1981). It has not been carried forward on a substantial scale, partly because of a declining interest in planning and the decline of operational research as a discipline; and partly because, at the time the problems were being specified, the requisite computing power was not available. The time is now ripe for a resurgence of operational research of this type and the first signs can be detected (see Birkin *et al.*, 1996).

This will lead to connections with another kind of research: the so-called NP-completeness problem. This is a class of problems for which an optimal solution cannot be found in finite computing time.

KEY IDEA 8.12

There are many research problems that can be formulated as design tasks – some of these as optimisation problems.

Under-developed systems analysis

INTRODUCTION

We have already noted above that there are many instances where potentially available models have not been developed in practice. The only areas where there is a lot of experience are transport and retail. We noted the number of groups still working on comprehensive models but these have not in general explored the dynamical properties of the models. We can chart out the research territory either in terms of system types or in terms of current problems. Two examples of the latter might be drugs and homelessness. A different kind of problems' perspective is represented by the shifting priorities of urban planners noted in Chapter 4: from an urban structure viewpoint to an economic development one, with cities competing with one another for inward investment. Here, we organise the argument that follows mainly in terms of system types and address their associated problem agendas within this framework. However, we should recognise that for new problems we need to define new composite systems.

The 'retail' model, with suitable adjustment and development, can be applied in other service areas, notably health and education. In the retail case, we might define the crucial array $\{S_{ij}{}^{mn}\}$ as the flow of spending by consumers of type m from residences in zone i to shops of type n in zone j. Recall that the associated model would be based on the residential distribution of spending power, $\{e_i{}^{mn}P_i{}^m\}$, and the attractiveness of shops in j for goods of type n, $\{W_j{}^n\}$. The flows would be inverse functions of the travel cost between i and j, c_{ij}. We begin with applications of this type to health and education, but we then proceed to address a broad research agenda, working through the whole range of classical models in their modern guises. The research agenda discussed in broad terms below can be expanded by referring back to the sectors enumerated in Tables 2.1–2.10 in Chapter 2.

HEALTH

In the health case, $e_i{}^{mn}$ could be interpreted as the demand for treatment of condition n from persons of type m; and $W_j{}^n$, a measure of the effectiveness ('attractiveness') of the treatment of n-type conditions. An obvious definition of $S_{ij}{}^{mn}$ then follows: it is the number of type m patients, resident in zone i, treated in hospital j for condition n. (Note in this case that j can be used to represent a particular hospital rather than a zone.) In practice, these definitions would have to be further refined, but a model based on this $\{S_{ij}{}^{mn}\}$ already contains some rich detail and some interesting research questions. We can also construct a variety of instructive performance indicators, in particular building on the argument regarding catchments in Chapter 6. In particular, we can measure the catchments of hospitals for a condition and person type:

$$\Pi_j^{mn} = \Sigma_i(S_{ij}^{mn}/S_{i*}^{mn})P_i^{mn} \qquad (8.1)$$

and then construct efficiency indicators such as W_j^n/Π_j^{*n}, if we take, in the first instance, W_j^n as a measure of provision. We can also calculate

$$\Omega_i^{mn} = \Sigma_j(S_{ij}^{mn}/S_{*j}^{mn})W_j^{mn} \qquad (8.2)$$

as a measure of the provision of n facilities to m-type people in zone i, and hence we can calculate a provision ratio Ω_i^{mn}/P_i^{mn} whose spatial variation (i) and class variation (m) will be particularly interesting.

The variation in $\{e_i^{mn}\}$, both spatially and by person type, will raise interesting issues in public health. If m is a social class index, for example, and n is a condition related to diet, then e_i^{mn} is likely to be greater for lower social classes – and for zones with a higher preponderance of lower social class households. The most interesting issues turn on the definition of W_j^n. So far, we have provisionally taken it as a measure of the scale of provision. For modelling purposes, if hospital consumers chose in the same way as retail consumers, then we would want W_j^n to measure effectiveness and attractiveness simultaneously. That is, we would want the consumer to be aware of the effectiveness measure and to use this as a measure of attractiveness. However, we know that the mechanism does not, on the whole, work this way in Britain, for example. Hospital referrals are through general practitioners; but then we could make the same assumption for them, acting on behalf of patients. Reality is different, however. The information available about effectiveness is at best imperfect, perhaps, for most practitioners, mainly anecdotal. In that case, it may well be that the W variable functions much more like its retail equivalent and can initially assumed to be measured by size.

One set of interesting research questions, therefore, is associated with behavioural model-building: how to define and construct the variables to achieve a reasonable representation of reality. An even more interesting set, however, is how to change both the system and associated behaviours to achieve a better outcome for patients. For instance, when provision ratios are calculated from current data, it will almost certainly be found that there are substantial variations and that a greater degree of equality would be better. This can be seen directly from studies that have been carried out on mortality rates for serious diseases such as cancer, where it has been discovered that there is tremendous variation (Birkin et al., 1996, chapter 6). In the case of cancer, this has been recognised officially through the Calman Report (Department of Health, 1996) which contains proposals to improve treatment regimes. Let us consider a hypothetical case of how modelling might help.

Suppose it can be shown that treatments are, on average, more successful if carried out in teaching hospitals for what we can define to be *tertiary* conditions. Consider a single disease and let n now represent type of hospital at j (say with $n = 1$ a teaching hospital, $n = 2$, a general hospital). The first stage of a project is to build a model that represents the existing situation, written formally as

$$S_{ij}^{mn} = S_{ij}^{mn}(e_i^m, P_i^m, W_j^n, c_{ij}) \qquad (8.3)$$

Let σ_j^{mn} be the survival rate of type m people in the type n hospital in j and let us assume that it is a function of the size of the hospital and the accessibility to teaching hospitals:

$$\sigma_j^{mn} = \sigma_j^{mn}[W_j^{(1)mn}, a_j^n] \qquad (8.4)$$

where a_j^n, using the measure introduced in Chapter 6, is a measure of accessibility:

$$a_j^n = \Sigma_k W_k^{(1)n}\exp(-\beta^n c_{ij}) \qquad (8.5)$$

We have picked out a factor of W_j^{mn}, defined as $W_j^{(1)mn}$, to measure size. (Of course, we have defined a substantial research problem in defining the functional forms of, and calibrating, models associated with equations (8.4) and (8.5).)

The basis of the model defined here lies in two hypotheses: that the larger the general hospital, the more likely it is to have the appropriate range of treatment regimes; and that if teaching hospitals are more accessible to general hospitals, then there is more likely to be a referral for tertiary conditions and hence a greater probability of successful treatment for the most difficult cases. Given the costs of building new hospitals, it is more likely that significant improvement can be achieved by changing behavioural patterns implied in the model – hence the suggestion made earlier to explore hub and spoke forms of organisation: in effect to make the teaching hospitals more accessible to general hospitals.

See Clarke and Wilson (1985a) for an early exposition of these ideas and chapter 6 of Birkin *et al*. (1996) for a more recent one.

EDUCATION

The education case can be addressed similarly: measuring schools according to 'league table' performance indicators relating to exam results leads to some schools being designated 'under-performing' schools. The underlying reasons are many and complex. We can provide a framework for analysis by building an interaction model, for example by defining an array $\{T_{ij}^{mn}\}$ where m again represents, say, social class of households in residential zone i and n represents a school type located at j. (As in the hospital case, j would represent the location of a particular school rather than a zone.) In formal terms, the model could be written:

$$T_{ij}^{mn} = T_{ij}^{mn}(e_i^{mn}, P_i^m, W_j^n, c_{ij}) \qquad (8.6)$$

For the sake of argument, suppose we restrict ourselves to secondary schools. The initial research task is to find effective ways of characterising m and n. It is likely that there are more subtle differences to be accounted for within the m label than traditionally defined social class; however, let us assume that a single label will suffice. In the case of secondary schools in the UK, for example, there are two distinct types: 'public' (which, of course, means 'private') and those mainly funded by the government); and those administered through local authorities. Further subcategories could be

defined. If, however, we take exam performance as a measure of success, then we have to begin by noting that published league tables show that this success varies tremendously within a 'type' of school. We would need to examine traditional factors such as average class size and we might then start to conjecture, for instance, that social class mix of pupils was a major factor.

The first major research task would be to characterise W_j^n as a measure which enabled us to predict the assignment of pupils from residential areas to schools. We would expect to build in a quality factor and our earlier preliminary broad analysis shows that it would be a difficult research task to construct this. We would then, as part of model calibration, have to represent the extent to which quality 'mattered' for different social groups. A particularly difficult research challenge would be an attempt to predict exam – and hence school – performance from a run of the model. If this *could* be achieved, then it would be possible, as in the hospital case, to address system provision issues and explore whether through some kind of re-arrangement – some kind of schools' hub and spoke (or supermarket), say – we could reasonably expect an improved performance, particularly for those pupils who were underperforming. This is an example of a particularly difficult research problem, but one that would be very rewarding. Whether a model could be fully calibrated or not, it is almost certain that interesting results would emerge. In that case, it would be an example of the model framework providing a basis for more effective *qualitative* analysis than might otherwise be possible.

It was argued in Wilson (1995) that these ideas could be applied to universities. Let the formal model equation be

$$T_{ij}^{mn} = T_{ij}^{mn}(e_i^{mn}, P_i^m, W_j^{mn}, c_{ij}) \qquad (8.7)$$

where we now interpret m as 'type of student' (say, for the sake of illustration, again social class, but the index itself could be a longer 'list'), n as 'university course' and j as 'university'. W_j^{mn} is now the attractiveness of university j for course n for students of type m. There are some interesting technical model-building questions: some courses, such as Medicine, will be supply-constrained; some, such as Italian, will be demand-constrained. So the model will have to be mixed – part attraction-constrained, part production-constrained.

Suppose the model can be built. It would then illustrate the application of catchment-based performance indicators such as those introduced in Chapter 6. Following the argument in Wilson (1995), we can define

$$X_i^{mn} = \Sigma_j T_{ij}^{mn}/T_{*j}^{mn}.D_j^{mn} \qquad (8.8)$$

where D_j^{mn}, in the usual way, is

$$D_j^{mn} = \Sigma_i T_{ij}^{mn} \qquad (8.9)$$

that is, the total number of type m students taking a type n course in university j. Define also

$$Y_j^{mn} = \Sigma_i T_{ij}^{mn}/T_{i*}^{mn}.P_i^{m} \qquad (8.10)$$

Using the argument of Chapter 6, we can see that X_i^{mn} is a measure of the effective delivery of the system to type m students resident in zone i and Y_j^{mn} is a measure of the catchment population of university j for type m students for course n. We can then, following what could be standard practice, calculate indicators such as

$$X_i^{mn}/P_i^{m} \qquad (8.11)$$
$$D_j^{mn}/Y_j^{mn} \qquad (8.12)$$

which are, respectively, the effective delivery per head of population by residential area and the university delivery per head of catchment population. These are to be contrasted with what are virtually meaningless indicators

$$D_i^{mn}/P_i^{m} \qquad (8.13)$$

and

$$D_j^{mn}/P_j^{m} \qquad (8.14)$$

If we find that X_i^{mn}/P_i^{m}, for example, is very low for Cornwall, then this gives a basis for solving the problem through some 'what if' explorations of alternative HE configurations.

SOCIAL SECURITY AND WELFARE

A contemporary problem referred to earlier was that of homelessness. How is this to be represented in a residential location model such as that shown in Chapter 6? Technically, the solution is straightforward: introduce a new class of 'residences' which indicates 'homeless'. It may be useful analytically to have a number of subclasses. These models represent people making 'choices' about their residential location. Few people probably choose to be homeless. Homelessness points to a failure of the social security or welfare systems, or to the failure to provide opportunities which enable more 'satisfactory' choices to be made. The interesting research topic, therefore, is not simply the modelling task, but its integration with a whole raft of social policy questions. It may then be that new concepts emerge and can be explored, such as that of *opportunity gaps* first introduced in Wilson (1972) but never developed.

AGRICULTURE

The first 'classic' model was that of von Thunen which he applied to agricultural land use. In its original form, it will appear very limited to the prospective contemporary user. However, we have shown in Chapter 7 that it can be generalised and made very powerful. This must open up new research opportunities: to apply these modelling principles to the modern forms of agriculture which have changed landscapes in a variety of economic and environmental contexts.

The formulation of the general model in a variety of specific contexts therefore forms the first set of research issues of this type. There are two other issues (of many) which can be cited as worthy of the modeller's attention. The first of these is to recognise the agribusiness system and the extent to which this determines agricultural land use as a component of it. The second is to explore the effects of reducing transport costs. The market gardening of the classic models now operates on a global scale. Fruit and vegetables which are not seasonal in one country can be flown in from another. This international trade could be modelled in an interesting way.

KEY IDEA 8.18

Key idea 8.18. We can continue to recognise the value of von Thunen's model. Great power can now be added to it and new models can be developed both of agricultural systems and of agribusiness. The study of international trade will make a particularly interesting research area.

UTILITIES AND COMMUNICATIONS

The following utilities 'flow': coal, oil, water, gas, electricity and information (with, to an extent, an historical progression implied by this list). All the utilities' industries will remain important, but it is the information sector that is growing fastest of all. As utilities flow between production sites and consumers, then the spatial interaction and location paradigm applies and the modelling techniques of Chapter 6 can be used.

Utilities typically have origins as resources at production sites and destinations represented by consumers at attraction sites. The reverse process, not included on the original list, but important and worthy of modelling effort, is that of waste disposal. Consumers can be both households and organisations. In many cases, there are various production steps which are carried out at different locations. These different kinds of flows can be modelled perfectly well within the paradigm.

The only sector that has been extensively modelled – with a history which originates in the 1960s – is water. For a recent example, see Birkin *et al.* (1996). The most exciting challenge is represented by the information sector. An obvious starting point is telephone communications. The *provision* in this case is unusual relative to the other utilities sectors: it is more like the transport system because it is the provider of the *network* that carries consumer–consumer interaction. The modelling task, as in the transport case, is to represent the interaction. Guldmann (1998) points out that this task has a long history, dating back to the paper by Hammer and Iklé (1957; cf. Iklé, 1954), who obviously worked with gravity models earlier. He himself estimates interaction models by using econometric methods, though using terms from intervening opportunities (Stouffer, 1940) and competing destinations approaches (Fotheringham, 1983). There still seems to be considerable scope for building spatial interaction models of the type presented in this book. The argument can then be taken a stage further and, in principle, the telecommunications' spatial structures could be modelled – the equivalents of shopping centres and networks could be modelled using the methods of Chapter 6.

What is particularly exciting about the sector at the present time arises from two aspects. First, new kinds of traffic are continually being invented, the most recent and striking example being the Internet. Secondly, new networks are continually being brought into play following deregulation (cf. Graham and Marvin, 1996).

This argument can then be extended to incorporate other kinds of information and information receiving and processing, in particular extending to all forms of computing and to the media. There are particularly important challenges arising from the convergence of the technologies which underpin telecommunications, computing and television.

KEY IDEA 8.19

Very little modelling has been undertaken of utilities' systems. Examples in relation to water systems show that there are many fruitful projects in these areas founded in interaction and location modelling methods. There are tremendous opportunities in the rapidly developing communications area.

MANUFACTURING

We remarked in the context of agriculture that the von Thunen model had not seen many contemporary applications. The same can be argued for the Weber model of industrial location, and for similar reasons: it was, in the main, a micro model applicable in local situations. We now have to think globally. However, the basic principles of interaction and location – coupled in this case with input–output analysis – still apply. It remains an interesting speculation whether the interaction paradigm will ever be fully combined with the input–output one, particularly in the context of industrial location (cf. Jin and Wilson, 1993). A whole series of research projects could be devised to articulate, through modelling, the structures of a variety of industries. Krugman (1996) has used an economic formulation of the industrial location model to formulate a new basis for a central place theory.

KEY IDEA 8.20

As we recognised von Thunen, so we should recognise the significance of Weber and the power we now have to tackle Weberian questions with the new modelling paradigm.

PROPERTY

An important sector in most economies is concerned with property and property development. It is often the property developers who determine land use by using their own methods to evaluate alternative uses of particular plots (or combinations of plots). What underpin these analyses are the rents that can be achieved for each alternative relative to the costs of different kinds of development. The prediction of rents lies at the heart of the land-use models that are developed within the interaction–location paradigm and there are undoubtedly gains to be achieved from research projects which link such model-based analysis to the needs of the property sector.

KEY IDEA 8.21

Property represents an under-developed sector and there are many research opportunities to be exploited.

MARKETING AND ADVERTISING

Other major industries are concerned with marketing and advertising. The enormous expenditure involved would justify the best possible analytical base. At present, the most sophisticated methods typically employed in relation to consumer goods use *geodemographics*: the analysis of populations, and hence markets, on a small area residential basis. What is not taken into account are the actual shopping patterns of the consumers as represented in the retail model. Another research challenge, therefore, with potentially enormous commercial pay-offs (both for those who use the new methods and for those who develop them) is to integrate the needs of the marketing and advertising sectors with the analytical power of the retail model.

KEY IDEA 8.22

Marketing and advertising represent another seriously under-developed sector with many research opportunities.

CULTURAL AND RECREATIONAL

Who could have predicted, as television 'took over' from the cinema, that there would in fact be an increase in the number of cinema screens? The new industry is, of course, different from the old: there are fewer cinemas, but they are larger ones with multiple screens. The transition is analogous to the shift to a mainly supermarket base in food retailing. An interesting research study would be to analyse in modelling terms the history of this. This argument can be extended to the whole of the leisure sector. In sport, for example, there is a rich supply of data (e.g. on attendances at football matches) and the model-based analysis of this would at least be worth a student thesis or two!

KEY IDEA 8.23

It can be argued that the cultural and recreational sectors represent an increasingly important part of the economy. Because many are 'consumer' based, there are many opportunities for model-based analysis and planning.

COMPOSITE SYSTEMS AND ALTERNATIVE FOCI

The illustrations in this chapter have been focused mainly on particular sectors. It remains only to remark that there are many ways of defining composite systems, most obviously in terms of the general urban model. For example, there is enormous scope for its application in historical geography and social and economic history, in providing a framework for the study of urban evolution over long periods. In the historical case, it is also possible to use the models to explore ancient geographical structures; see Rihll and Wilson (1987a, b, 1991) for an application of the singly-constrained spatial interaction model predicting equilibrium structures using only point data on known sites in Ancient Greece.

KEY IDEA 8.24

There are many possible applications in history, archaeology and historical geography.

It is also possible to define different kinds of foci. We consider in turn the movement of people, ethnic groups and networks.

If we consider the movement of people between relatively large areas, then this will be considered as migration rather than residential (re)location. Interaction models can still be used – and these have typically been developed within demography. There are some major research problems, for example in considering the major movements across and between continents of populations in recent times. See Rees (1996) for a European example.

It is also appropriate for some purposes to focus on ethnic groups, particularly, to consider the evolution of their spatial structures in cities. Rees and Phillips (1996) provide a recent example.

Little research has been done on the evolution of networks; or, to take a very different kind of example, on the modelling of markets. The study of network evolution has a long history (e.g. Haggett and Chorley, 1969; Wilson, 1983b). There is an obvious interest not only in network evolution as such, but the extent to which network investment can drive economic development (Gillen, 1996).

KEY IDEA 8.25

The model-based approach involves a system of methods – and composite systems and models will be necessary to meet particular objectives and cases ranging from migration studies, foci on ethnic groups to the study of network evolution.

Ongoing research challenges

INTRODUCTION

The research opportunities described to this point in this chapter could be implemented immediately, and much of it is within reach of student theses. It is worth spending some time, however, on the longer run to see whether we can plausibly glimpse any of this future. We discuss in turn the conceptual, mathematical and computational challenges which underpin this agenda. These ideas are pursued in more detail in Wilson (1999).

CONCEPTUAL CHALLENGES

The whole approach of this book has been essentially interdisciplinary. A long-term challenge is to explore the extent to which full interdisciplinary integration can extend the power of urban and regional modelling. It is clear that there are many valuable analyses and applications which can be undertaken on a subsystem or disciplinary basis, but judgements will have to be made, in the light of further experience, about the extent to which full integration is necessary. At the moment, of course, it is not achievable! Insufficient thinking power is being applied to the interdisciplinary opportunity.

KEY IDEA 8.26

It would be a major (but exciting!) task for a large research group to see how a fully integrated interdisciplinary approach to urban and regional modelling could drive theoretical development forward. This is a precursor for the 'big science' argument below.

A similar argument can be applied to the benefits of integrating the insights of different approaches which themselves lie outside disciplines. The concepts of time geography provide one example; the work begun by Becker (1965) on time consumption is a related one. These have not been fully incorporated into the modelling canon. The bulk of the most recent work has been in the transport field (e.g. Axhausen and Garling, 1992; Fox, 1995).

There are specific challenges. Arising out of economics, for example, we need to explore the full integration of prices and rents. It is a feature of the development of the field that where new opportunities are recognised within a discipline, as with Arthur's contributions on positive returns to scale and path-dependence in economic analysis, the applications generated within urban analysis use models which are obviously inadequate. Conceptual integration would avoid this.

In these contexts, it is worth speculating that the ultimate development of complexity theory will have the same kind of broad impact that was achieved from the development of nonlinear dynamics in the 1970s. When it becomes possible to articulate in mathematical (or computer simulation) form such concepts as *emergence*, and the urban equivalents of *learning behaviour* in artificial intelligence, then significant advances are likely in urban modelling as will be the case in many other fields (cf. Holland, 1995, 1998). Note that in general we are seeking models through complexity theory of structures whose manifestations may be relatively simple! This is not an argument about chaos (cf. Cohen and Stewart, 1994; Williams, 1997).

It can be envisaged that huge advances are possible in the use of models in planning and design – in part because there has been so little experience, but also because we can expect associated advances in the associated computer science.

MATHEMATICAL CHALLENGES

We argued in the early chapters of this book that the form of spatial representation, and in particular, the use of a discrete zoning system rather than a continuous space representation, is important in facilitating the representation of models and that the mathematics of the discrete zone representation is 'easier', and therefore more powerful. We have already had hints that this argument might be extended to the mathematical representation of models more directly. For example, the micro simulation representation is one alternative (though the advantages in this case are perhaps mainly computational); and we saw in Chapter 6 that, by representing some entropy-maximising models in mathematical programming form, we could use some of the theorems and insights of the latter formulation to offer new results and interpretations not immediately clear in the former one. It is part of the richness of mathematics that this can be done. Ultimately, it will be possible to see such alternative representations as subcategories within some more powerfully represented single branch of mathematics – and probably for a good enough mathematician, this can be done now! A fruitful question to pursue in the following subsections is: are there any more correspondences of this kind, and new insights to be gained? We will see the beginnings of positive answers. This will help us to draw together a variety of approaches to understanding the equilibria, and particularly the multiple equilibria and the change from one kind of structure to another at critical points.

We have had hints and conjectures earlier (in Chapters 4 and 6, and earlier in this chapter) and we can now draw these together and extend them. The possibilities to be discussed are as follows:

- the use of an alternative entropy function
- fixed-point theorems
- Cauchy's theorem: the complex plane as a new representation

- Markov processes
- game theory
- neural networks
- cellular automata

It is useful, taking the retail or service model as an archetype as usual, to think of the model components as they represent consumers and retailers. In this book, we have essentially used entropy-maximising models of consumer behaviour (and recognised a number of equivalent representations) and we have built systems of simultaneous nonlinear difference equations to represent retailer dynamics. The two systems of course interact, albeit with fast and slow dynamics respectively. We noted that a particularly difficult problem was that in trying to locate equilibrium solutions to the retailer difference equations, the analysis for a particular location depended on the configuration represented by the choices of retailers at all other locations – the *configuration problem*. We can now examine, as part of our definition of the ongoing research agenda, what intuition suggests about the contributions of these new mathematical representations. We consider each of the six possibilities noted above in turn, though we can also add that there is a common thread: each approach usually has a *Bayesian* version, to represent uncertainty or probabilistic behaviour and, of course, as we saw in Chapter 6, this connects to one of the representations of the entropy-maximising method, which is one clue to the basis of some of the links.

Dynamics and Fisher information

We begin with a speculative avenue of research which does not involve a wholesale new mathematical representation but which involves a *new representation of the entropy function* that lies at the core of our models. Frieden (1998) has recently argued that the entropy function that lies at the core of our models, the Boltzmann-Shannon entropy, $-\Sigma T_{ij}\log T_{ij}$, can, in the context of physics, be effectively and powerfully replaced by Fisher information – essentially a sum of squares, $\Sigma(T_{ij} - T_{ij}^{obs})^2$. He argues that this is a better measure of information. But, more importantly, he argues that this measure leads to *dynamic* models while the Boltzmann-Shannon measure does not.

In the special case of thermodynamics, not surprisingly, the two can be made equivalent – but not in general. The research task, therefore, is to rewrite the core models in terms of Fisher information and to see whether this will lead to alternative dynamic models, and an alternative (and perhaps improved) theoretical base.

Approaches to equilibrium: fixed-point theorems

The first step in understanding the dynamics of spatial structure is to understand the possible underlying equilibrium states. We mainly used an *ad hoc* approach in Chapter 6 which did provide considerable insight. A research challenge is formally to integrate the mathematical theorems which can be brought to bear: Brouwer's (1910) fixed-point theorem and those associated with nonlinear mathematical programming have already been referred to earlier (cf. Scarf, 1973a, b, for an economist's approach to this area, and see Casti, 1996, chapter 2, for a general discussion of this theorem). We will see in the paragraphs below that further insights can be obtained using the theorems of Markov processes and game theory. In each of these cases, the models need to be represented in the appropriate form so that the theorems can be applied.

Cauchy's theorem: the complex plane as a new representation?

One additional thought can be introduced at this stage: whether Cauchy's theorem has any relevance (see Osborne (1999) for a clear statement of the mathematics). The basis of this mathematical challenge is as follows. Cauchy's theorem is concerned with identifying the singularities of a function of a complex variable in a plane. The spatial positions of these singularities could represent urban spatial structure – say the positions of shopping centres. If, therefore, we could represent the dynamics, or the equilibrium equations, in terms of a function of a complex variable, with the complex plane representing geographical space, then perhaps Cauchy's theorem could be brought to bear? In principle, the positions of singularities could be identified in a way that avoids the configuration problem.

Markov processes

In the application of *Markov theory* by Smith and Hsieh (1997), the focus is on the consumer (in their case, the migrants in an urban system) and the probabilities of moving are represented in a Markov framework. This does allow the general properties of Markov theory to be applied and in particular to give further insights into equilibrium states – as existence theorems on stable *steady states*. This result turns on Brouwer's fixed-point theorem which is applied to good effect in so many of these different representations. Interestingly, since we are continually interested in alternative model representations, we note that they also present an equivalent mathematical programming formulation. We have already suggested above that this helps us in principle with aspects of configurational analysis, demonstrating that for a very complex dynamical system – the retail system – in which it is not possible to solve the equations explicitly, stable configurations do exist. There is an exercise to be done to show that the migration model used by Smith and Hsieh can be transformed into a retail model.

KEY IDEA 8.29

There are deep mathematical problems still to be tackled: one, for example, arises because the current analysis of the dynamics of one zone depends on the assumed fixed configuration of all other zones.

We noted in Chapter 6 that Arthur (1994a) had developed a model of industrial location based on a Markov process which combined geographical advantage with increasing returns to scale. In our notation, we could take W_j as a number of firms and his model is then based on the probability that the next firm locates at j. This probability is based on the assumed benefits of locating at j which take the form $q_j + g(W_j)$, where q_j represents the geographical advantage and g the agglomeration economies. In some ways, this is not an interesting geographical model; it will be inferior in representing behaviour to the models used in this book, which also, as we have seen, incorporate increasing returns to scale. However, the underlying mathematics, as in the case of Smith and Hsieh above, may offer new insights. Arthur considers the problem of probabilistically locating the 'next firm' in a probabilistic Markov process as equivalent to a so-called 'urn' problem (drawing coloured balls from an urn) and he, with two colleagues, has extended the mathematics of this area to show what happens when there are increasing returns to scale (Arthur *et al.*, 1983). An interesting research problem would be to compare the results of this approach with those of others described here which are concerned with the properties of equilibria in nonlinear systems. A further research task would be to seek to represent the core interaction–location model in this format.

Game theory

Next, we turn to *game theory* (guided by Gibbons, 1992). The focus here should be on the retailer (or the developer or planner, since essentially we are assuming that there is one retailer at each location who is determining the size of the retail facility there). We assume that consumer behaviour is governed by a standard interaction model. If there are N locations, then the single retailer at each location can be considered to be a player in an N-person game. From one time period to the next, each retailer will decide whether to change the size of the facility at that location, in each case making a guess about the strategy to be adopted by the other players. There are three possible beneficial consequences of this idea. First, it shows explicitly how one retailer is dependent on what he or she cannot know (i.e. the strategies of the other players) and this precisely coincides with the problem of what we have been calling *configurational analysis*. Secondly, in this case, there is an important general theorem: Nash (1950) showed, in a famous Nobel prize-winning paper, that, under very general conditions, an equilibrium would exist. (This is almost certainly on the same basis as that shown in the Markov representation as it too depends on Brouwer's fixed-point theorem.) Thirdly, there is a Bayesian form of this problem, representing retailer uncertainty, and again, the equilibrium still exists. This presumably coincides with the attempt in Chapter 6 to build this into the dynamic model used in this book. It is worth noting that in any complicated game situation, Nash showed that, typically, there will be multiple equilibria, and this, of course coincides with our own experience.

Neural networks

We have already seen that *neural network methods* have been used for the calibration of standard interaction models, and to generate new models – essentially through perceiving the networks as a statistical machine. In this last case, the interaction models are given a very general formulation and represented in a neural network which then 'learns' how to generate a best-fit model. This, as we suggested earlier, may provide clues about new models, but does not seem directly to add to the theory. The major challenge is to see whether new models can be developed within this framework which can in some sense represent learning mechanisms for complex spatial systems; and whose behaviour will then reveal new forms of emergent structures. What we have to recognise, however, is that any model can be represented as a neural network, and as another representation of this general class, building on a powerful result of von Neumann, any model can be represented as a set of cellular automata (cf. Holland, 1995; Hillis, 1987; von Neumann and Morgenstern, 1947). We review these models briefly in the next subsection.

Cellular automata

At this point, we do begin to see glimpses of tantalising connections: *cellular automata* are usually represented on square grids, e.g. as a set of rules for determining whether a particular square on the grid is 1 or 0 at a particular time – a specification of a dynamics model. If one thinks of the 1's as being shaded, then this generates a sequence of patterns as in the retail dynamics model of Chapter 6. We could, in principle, therefore, invoking von Neumann's general theorem, represent the dynamic retail model in cellular automata form. Are there any general theorems which then might help us? Not to the author's knowledge at the present time, but there are interesting hints in one of Holland's (1998) cellular automata models, for which he shows that

a very simple set of rules will generate patterns – his 'gliders' – which, in form, are stable over time, and it is this kind of result that will help us in urban modelling. However, his discoveries were made from repeated simulations rather than being derived mathematically – the point of the exercise being to show that simple automata rules would not usually generate stability (as the *emergence* of forms), but some might.

If we are looking for analogues from entirely different directions, there may be one with the *N*-body problem in physics: simultaneous equations to which there are no analytical solutions. Recently, Buck (1998) has shown that there are particular stable solutions, e.g. when some of the bodies are in particular orbits. Interestingly, these can be derived in situations where the bodies are in particular initial configurations which are found by symmetry arguments. There is almost certainly a general class of problems here which occur in a number of disciplines in different representations.

KEY IDEA 8.30

There is much opportunity for deeper mathematically based research: (i) to understand how the mathematics generated from different perspectives can be integrated; (ii) to generate new insights by deploying theorems and interpretations which are available in different branches of mathematics.

COMPUTATIONAL CHALLENGES

It will already be clear from the previous section that it is difficult to separate the mathematics from the computer science: a number of the mathematical challenges are rooted in high-speed computers, particularly the modelling methods associated with neural networks and cellular automata, but also those which bring human interpretation into the picture through the visualisation of results on computer screens. We discuss briefly in turn the research end of the spectrum relating to concepts that were introduced briefly as part of the tool-kit in Chapter 4 and again, in slightly more detail, earlier in this chapter. They are algorithmic thinking, representations, visualisation and simulation. All have played a role in the history of modelling complex spatial systems. Here, we focus on the longer term challenges.

There is no doubt that *algorithmic thinking* can offer new avenues of model-building – the references to neural networks and cellular automata above illustrate this. It can also trivialise some complex problems – as in the *systems dynamics'* school of model-building. The research challenge is therefore perhaps twofold: to see whether we can invent new forms of plausible and realistic models; and secondly, to see if we can integrate the kind of analysis that can be deployed with mathematical tools with what computer modelling offers. Theory (e.g. Turing or von Neumann) indicates that this should be possible, but it is rarely attempted with models that have been developed as computer algorithms.

Where the adoption of a computing or algorithmic perspective may be particularly effective is in exploring different kinds of *representation*. Microsimulation, for example, is rooted in a computer algorithm. (Interestingly, in that case, it is possible to be explicit about the mathematics represented by the algorithm, because it is a computer representation of a known mathematical model.) We have seen that it should be possible to integrate the ideas of object-oriented programming with computer modelling, and in that case, while the way in which a real system may be represented more directly than in a mathematical model, the link to mathematical interpretation may be more difficult to achieve.

Intuition suggests that there is much progress to be made in developing the *visualisation* benefits of computer modelling. For example, we are used to seeing visual portrayals of dynamics for relatively simple mathematical systems on the one hand, and model outputs – of computer models representing more complex systems – on the other. The former suggests that we should be able to add more analytical power to our interpretations of the latter.

Finally, there is the issue of *real-system simulation*. Gates (1999) has drawn attention, with many others, to the benefits of *data mining*. At present, the outputs of data mining are as likely

as not to be inputs into relatively simple statistical analyses. What we have seen in the arguments of this book is the possibility of using mined data to calibrate quite sophisticated models and thus to add analytical capability.

Integration and implementation

THE CHALLENGE OF INTEGRATION

It has been emphasised throughout the book that we have a constant challenge of integration: there are many different approaches to the same modelling problems and attempts to integrate concepts and different mathematical approaches, for example, are nearly always fruitful. Because the argument runs so deeply, it is not helpful to try to summarise it; rather simply to note formally the dimensions of choice in terms of the research agenda as discussed in this chapter:

- applications areas
- conceptual
- mathematical
- computational

This defines a high dimensional combinatorial space and the clever researcher will be able to identify the fruitful nodes in this space. Many pointers have been provided in the book.

IMPLEMENTATION

It was argued at the outset of this book that understanding cities and regions represented one of the major scientific challenges of our time. I have long argued that it deserved the status and funding of 'big science'. If this was possible, then major databases would be systematically and easily available. The models themselves can fill in gaps, e.g. through microsimulation. These would link to intelligent geographical information systems – IGISs that would provide twenty-first century atlases. It would be relatively straightforward to extract particular systems for urban planning, retailing, health, education, fire and police services as described in this chapter. The benefits and scale economies would be enormous. We have also seen in the later part of this chapter that there is scope for a substantial basic research programme.

What kind of organisation could deliver this future? Governments would have to play a key role in the early stages, but there could be a variety of agents and funders. We have to bear in mind, of course, that there have been many famous failures in large-system development, but the increasing cheapness and power of the technology should reduce this risk in the future. Any country that can succeed in this enterprise could then find that it had a new major export industry!

Appendix 1
The basic notation and some examples

For many purposes, the most appropriate 'picture' of geographical systems is based on discrete zone systems and it is worth developing the notation for those who are not familiar with it. We can also take this opportunity to develop some ideas which arise from this notation.

First, we should remark that there is nothing unfamiliar about discrete zone systems: any map of sub-area units, say local authority wards, is such a system; so is a kilometre grid square system used for the presentation of many maps. The data associated with these systems are also familiar: tables of population and employment by ward, for example. However, it remains less familiar to progress to one higher level of abstraction: to describe a system based on discrete zones by subscripted and superscripted algebraic variables. Provided any reader can achieve this particular step, the description of the evolution of geographical theory can be followed in a straightforward way. If there is a difficulty, it usually arises in the switch from names of individual zones on a map to consecutive integers, 1, 2, 3, ... N say (compare Figure A1.1a and b in this respect). The numbers are just alternative names, of course, but are more convenient for algebraic description.

One further step remains to be climbed on the abstraction ladder. It is inconvenient to have to name every flow separately: $T_{1,23}$ as the flow from zone 1 to zone 23; $T_{5,36}$ as the flow from zone 5 to 36; and so on. It is therefore useful to be able to label a *typical* zone as i or j and to speak of T_{ij} as the flow from zone i to zone j. If we put $i=1$, $j=23$; or $i=5$, $j=36$, we recover our earlier specific cases. Figure A1.1c shows such a pair of zones picked out; but now they can be *any* pair of zones of Figure A1.1b.

This begins to give a more powerful notation in another sense. We can use the matrix notation $\{T_{ij}\} = \mathbf{T}$ to denote the whole *set* of flows for $i = 1, 2, 3 \ldots N$; $j = 1, 2, 3 \ldots N$. We can write down one equation for T_{ij} instead of 10,000 (for $N = 100$, say) separate equations for each individual flow.

To fix ideas further, consider a particular subsystem which is used as an example in a variety of contexts: the retail system. This involves a set of consumers, a set of shops or shopping centres, and flows between them. Suppose for simplicity, we work with aggregates and do not distinguish different kinds of goods – or, equivalently, that we are dealing with a single type of good. Then, let e_i be the per capita expenditure by residents of zone i, P_i, the population of zone i, and then e_iP_i is the total expenditure from i; let W_j be the amount of retail floorspace in zone j; and S_{ij}, the flow of expenditure on retail goods from i to j. If we had observations for a five-zone system, we could plot the information on a map as shown in Figure A1.2 or it could be tabulated. Either mode of presentation is normal and familiar. With a little practice, it is easy to see that the algebraic notation is simply a convenient shorthand for something that is familiar anyway. The algebraic notation is displayed in Figure A1.3 (where we have added D_j, the total inflow into zone j). We have added T_{ij} as defined earlier to the diagram, and C_{ij}, which will be used later as the (generalised) cost of travel between i and j. The reader is assumed to be familiar with summation signs, so that

$$\Sigma_j S_{ij} = e_i P_i$$

and

$$\Sigma_i S_{ij} = D_j$$

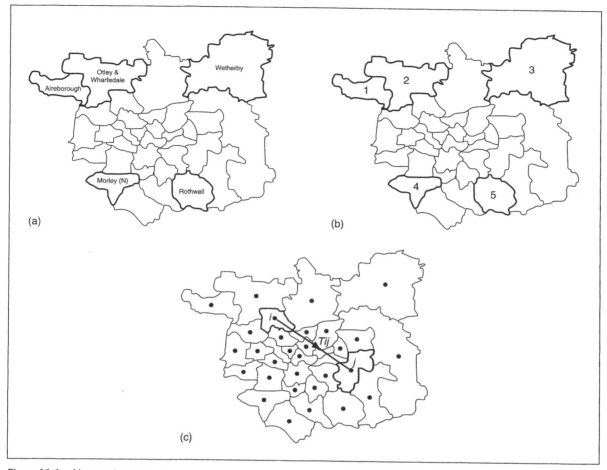

Figure A1.1 Maps and zone numbers

Most of the resulting models can be understood, at least in outline, by the use of elementary algebra with these kinds of variables. However, one other piece of mathematical equipment is needed: since we are going to be concerned with dynamical models, it is useful to understand the mathematical notation for rate of change. We can introduce this in relation to the variables defined above. Take P_i, which we defined as the population of zone i. Suppose we now add a label, t, to denote time, so $P_i(t)$ is the population in zone i at time t. Then

$$\delta P_i(t, t+1) = P_i(t+1) - P_i(t) \qquad (A1.1)$$

is the change in $P_i(t)$ from time t to time $t+1$. The symbol δ means 'change in', and labels $(t, t+1)$ are added for clarity. Sometimes the upper case symbol Δ is used instead of δ. Also δt (or Δt) may be used to denote 'an increment in time t', and so it is possible to define the time interval $(t, t + \delta t)$. With the previous definition, in equation (A1.1), we took $\delta t = 1$.

Another step is then to hypothesise that the rate of change in a variable is a function of the variable itself or other variables. For example, we might have

$$\delta P_i(t, t+1) = aP_i(t) \qquad (A1.2)$$

where a is the net birth rate in zone i (births per head of population minus deaths per head of population). If a is positive, a sequence of calculations using equation (A1.2) leads eventually to an infinitely large population – Malthusian

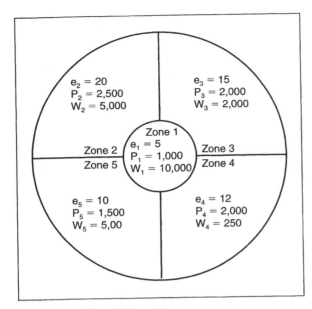

Figure A1.2 Mapped information

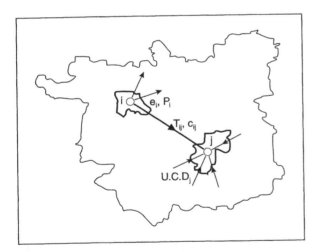

Figure A1.3 Spatial interaction variables

growth (cf. Figure A1.4a). It is common then to assume that a_i is modified by a factor $(1 - P_i/K_i)$ – to represent a declining net birth rate as the population grows, to represent the limiting effects of resource capacity or whatever. Then

$$\delta P_i(t, t+1) = a[1 - P_i(t)/K_i]P_i(t) \qquad (A1.3)$$

Then if $P_i(t)$ exceeds K_i then $\delta P_i(t)$ becomes negative. So for obvious reasons, K_i is known as the 'carrying capacity' of i. Note also that

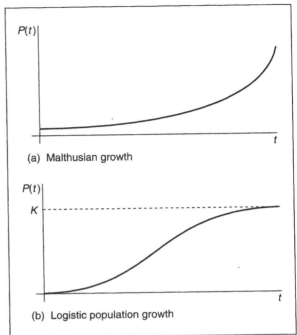

(a) Malthusian growth

(b) Logistic population growth

Figure A1.4 Malthusian and logistic growth

$$\delta P_i(t, t+1) = 0 \qquad (A1.4)$$

that is, there is no change when

$$P_i(t) = K_i \qquad (A1.5)$$

(because, then, $1 - P_i(t)/K_i = 0$). So $P_i(t)$ remains equal to K_i for future values of t. K_i is known as an *equilibrium value* of $P_i(t)$. Typically, for an equation specifying change like (A1.3), $P_i(t)$ will initially not be in equilibrium but will approach it. This can easily be tried with simple numerical examples. This is known as logistic growth (see Figure A1.4b).

An equation like (A1.3) is called a *difference equation* and it relates to discrete time intervals. It is sometimes appropriate (though as with space, often less convenient in practice!) to use continuous time, and a notation is needed which represents rate of change at a point in time. Suppose we rewrite the definition (A1.1) in terms of an increment in time, δt:

$$P_i(t, t+\delta t) = P_i(t+\delta t) - P_i(t) \qquad (A1.6)$$

The rate of change per unit of time (which we had directly in (A1.3) because of the t to $t+1$ basis) is now

$$\delta P_i(t)/\delta t = [P_i(t + \delta t) - P_i(t)]/\delta t \qquad (A1.7)$$

Consider now what happens in the limit as $\delta t \to 0$. The right-hand side of the equation tends to $0/0$ which is meaningless. However, it turns out that it is usually possible to calculate appropriate finite values for this limit which are denoted by

$$dP_i(t)/dt \qquad (A1.8)$$

As with δ, d/dt should be thought of as a complete symbol denoting 'rate of change at a point in time'.

If $P_i(t)$ is a function of variables other than t, then the symbol ∂ may replace d to indicate that there could also be rates of change with respect to other variables and this one is partial. But this is a refinement which is not considered further in the present context.

$dP_i(t)/dt$ is known as the *derivative* of $P_i(t)$ with respect to t. An equation involving a derivative is known as a differential equation. An example would be the continuous time analogue of (A1.2):

$$dP_i(t)/dt = a \qquad (A1.9)$$

The important conceptual point to grasp is that hypotheses and theories can be stated in terms of rates of change, and then a natural mode of formulation is in terms of difference or differential equations.

It is useful to conclude this introduction with a comparison of the discrete zone system and a continuous-space one. In algebraic terms, we could take (x,y) as the Cartesian co-ordinates of a point and $P(x,y)$ as the population density at that point. To obtain a population, we have to integrate; and so make greater demands on mathematical knowledge. More importantly, even with this knowledge, problems are much more difficult to handle than in the discrete case; and often impossible.

There are some losses of course. The spatial level of resolution is more coarse with a discrete zone system. But questions about boundaries, for example, can still be tackled. It is a matter of whether a zone is 'inside' or 'outside' a particular boundary. A comparison of the two systems is shown in Figure 3.1.

Appendix 2
Geography's classical theorists

Introduction

Every discipline has authors who are commonly cited as being responsible for its foundations. In this appendix, we present the work of those authors (not always geographers) whose work has generated much of the underlying theory of geography – and hence the modelling foundations of urban and regional analysis. They are still referred to in most textbooks and the presentation of their ideas in such contexts is often both sketchy and uncritical. Here, we try to offer a presentation that is better than sketchy, though inevitably it still has its limitations. The aim is to expose the main assumptions as well as the main ideas, and thus to provide a complement to Chapter 5. We discuss in turn von Thunen and agricultural land use; Weber and industrial location; Palander, Hoover and Hotelling, who worked on the identification of market areas and the problem of competing firms; Christaller and Losch, and central place theory; the urban ecologists, Burgess, Hoyt and Harris and Ullman; and the originators of the gravity model and the geographers' study of spatial interaction.

Agricultural land use: von Thunen

INTRODUCTION

Von Thunen is rightly praised by many commentators on location theory for the importance of his contribution; for the generality of some of his concepts; for the power of his theory; for the associated empirical work; and for being so many years ahead of the theorists who followed him. He was an economist and a practising farmer, and so it was not surprising that his main interest was agricultural location. He was also responsible, probably from his interest as an employer, for a theory of the wage rate.

His most famous book, *The Isolated State* (von Thunen, 1826), includes a development of a theory of agricultural location based on an enormous amount of empirical data. Although he has been described by economists as the first econometrician, his style of work was essentially hypothetico-deductive. He introduced simplifying assumptions in order to be able to get to grips with the problem which interested him; and he was then prepared to relax these in various ways. He was the first model-builder in location theory.

His work was based on the concept of 'economic rent', or 'location rent', and this has had an influence far beyond the study of agricultural location. In the next section, we present his main results first for two different problems at the meso scale, i.e. intensity of production and land use; and then shift to the micro scale, i.e. to the village or farm.

THE MAIN ELEMENTS OF VON THUNEN'S THEORY

Von Thunen's own presentation of his theory is sometimes difficult to follow because, on the whole, he uses the arithmetic of many case studies to make his general points rather than the greater power of relatively simple algebra. Here, we use the algebraic summary of his methods which has been used by most other commentators, particularly following the work of Dunn (1954). This also serves the very useful purpose of showing the reader who may not be familiar with formal modelling how very powerful results can be obtained on the basis of very simple algebraic and geometrical arguments.

Von Thunen's 'isolated state' consisted of uniformly fertile agricultural land surrounding a single market town. Here we use Dunn's notation – though in Chapter 5 we amended this slightly to bring it into line with the general notation used in other chapters. For some crop, not yet specified, let Y be the yield per unit area, p the price per unit weight, and r the transport costs per unit weight per unit distance from the market. Then the total revenue which can be obtained per unit area cultivated is Yp; the production costs for this total, Yc; and the transport costs for a farm at distance d from the town total, Yrd. The surplus potentially available to the farmer developing this crop is therefore given by E in the following formula:

$$E = Y(p - c - rd) \qquad (A2.1)$$

In addition to assuming uniform fertility (which means that Y can be treated as a constant), we also assume uniform labour and other production costs and uniform transport costs which means that c and r can be taken as constants. Also, the market price is taken as fixed (p) and it is assumed that there is a market for any amount of the crop which is produced (which in economic terms means assuming an infinite elasticity of demand).

E is essentially a measure of potential profit. It is known as a 'rent' not because this is necessarily the amount that the farmer would have to pay as rent, but because it is the maximum amount he would be prepared to bid as rent for the land to use for that crop at that distance from the market. The owner of the land, in a perfectly competitive market would, of course, be able to extract this amount from a tenant farmer as rent in the colloquial, rather than economic, sense. An implicit assumption is therefore also made here that the farmer should include the price of his own labour, and what might be called 'normal profits' as part of his *production* costs.

Since all terms in equation (A2.1) except E and d, are being treated for the present as constants, it can be written as

$$E = a - bd \qquad (A2.2)$$

where

$$a = Y(p - c) \qquad (A2.3)$$

and

$$b = Yr \qquad (A2.4)$$

and a and b are, of course, constants.

If we aim to plot the relationship between E and d geometrically, then the form (A2.2) shows it to be a backward-sloping straight line with slope $-b$ (Yr) and intercept on the E axis of a [$Y(p - c)$]. This last quantity is the maximum rent which occurs at the market itself. The plot is shown as Figure A2.1. This shows one new feature immediately: there will be a distance, marked as d^0, beyond which it will not be profitable to produce that particular crop. d^0 is, of course, the intercept on the d-axis of the straight line and its value is $(p - c)/r$. This is a result that will be useful later.

The concepts introduced so far provide the basis for exploring von Thunen's two main results at the meso scale. He first considered the *intensity* of production of a crop. This is defined by varying the amounts of inputs to the farming process and they are measured in terms of production costs, c. By increasing the input of labour or fertilisers, for example, the yield (Y) can be increased. We are now beginning to move away from treating c as a constant and exploring what happens when it varies. In effect, a functional relationship is anticipated between Y and c which shows Y increasing with c. This can be written

$$Y = Y(c) \qquad (A2.5)$$

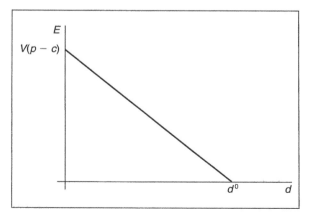

Figure A2.1 Rent versus distance from market

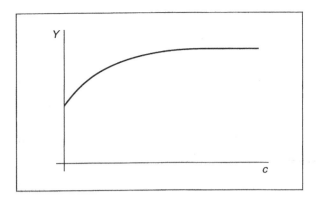

Figure A2.2   Yield versus intensity of production

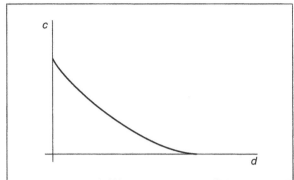

Figure A2.3   Intensity of production versus distance

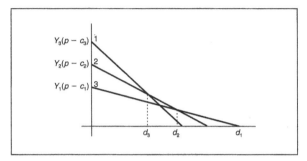

Figure A2.4   Rent versus distance for different intensities of production

in a formal sense. Geometrically, it can be expected to take the form shown in Figure A2.2: the increases in Y lessen after a time from the *law of diminishing returns*. However, it does mean that, nearer the market, it may pay the farmer to increase his inputs and hence yields because the increase in revenue so gained may outweigh the corresponding increase in transportation costs. Whether this is the case or not depends on the shape of the curve plotted in Figure A2.2 in any particular case, and on the relative sizes of production and transport costs. The broad result, however, is clear: other things being equal, we would expect to find more intensive cultivation of a crop nearer the market, and that the intensity of cultivation will decline with distance from the market until the point is reached, already seen in Figure A2.1, at which it is not profitable to cultivate the crop at all. The level of intensity, as measured by c, could then be related to distance as indicated in Figure A2.3. Note that there is no reason to expect that the relationships in Figures A2.2 and A2.3 should be linear.

The argument so far has been conducted on the basis that it is possible to vary the intensity of farming, measured by the input levels, continuously. In many cases, this will not be so. As was observed by von Thunen and many since, the farmer may be choosing between systems: two-, three- or four-cycle rotations for example. It may be that, between given systems, there are differing levels of inputs, but that within a system, c is constant. In this case, Y and c can be considered fixed for a system, and so Figure A2.1 applies for each system, but with different values of Y and c (and hence several straight lines of the form shown superimposed in Figure A2.4).

Figure A2.4 shows three systems, and it has been assumed that by increasing c from c_1 to c_3, the yields increase to such an extent that

$$Y_1(p - c_1) < Y_2(p - c_2) < Y_3(p - c_3) \qquad (A2.6)$$

as shown geometrically by the intercepts of the figure. In this case, the highest rent is obtained by using system 3 from $d = 0$ to $d = d_3$, system 2 from $d = d_3$ to $d = d_2$, and system 1 from $d = d_2$ to $d = d_1$. For $d > d_1$ there will be no cultivation of any crop. In this case, therefore, the land-use pattern exhibits three rings of decreasing intensity of cultivation away from the market, as shown in Figure A2.5. We should, however, remember that this can only occur if a particular relationship is assumed between Y and c for the three different systems: essentially Y_i ($i = 1, 2, 3$) has to increase more rapidly with c than the factor $(p - c_i)$ decreases.

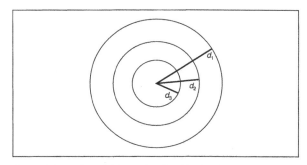

Figure A2.5   Rings of intensity of cultivation for a single crop

The relationships between the gradients of the lines follow more straightforwardly: as c increases, Y increases, and therefore the slope Yr increases, and this is the result that has been used to construct the figure.

We can now turn to the second result, which is usually considered the more important of the two, and is sometimes the only one presented in textbooks. We can make direct use of the concepts that have been set up above. Now, instead of considering varying intensities for a crop, we consider alternative crops; or more generally, alternative agricultural uses, including, for example, grazing for meat and dairy products (though this includes complications of 'joint costs' which we will neglect for the time being – by assuming the products of grazing form a composite good).

In order to proceed explicitly, we add the subscript k to the various terms of equation (A2.1) but otherwise let all the definitions stand for what now becomes 'mode of use' k. The equation can now be written

$$E_k = Y_k(P_k - c_k - r_k d) \qquad (A2.7)$$

There is no need, of course, to add a subscript to the distance term. This can also be written, from equation (A2.2), as

$$E_k = a_k - b_k d \qquad (A2.8)$$

with obvious definitions for a_k and b_k.

We now have a series of backward-sloping straight lines for each use k. We use an elegant notational device due to Stevens (1968) to derive von Thunen's main result explicitly, though the derivation can also be followed in verbal and geometrical terms. First, order the uses so that

$$Y_1 r_1 > Y_2 r_2 > Y_3 r_3 > Y_4 r_4 > \dots \qquad (A2.9)$$

This means that the crops are ordered in terms of decreasing transport cost gradient per unit distance: the first has the steepest gradient and therefore we expect it to be produced nearest to the market centre. As a preliminary, however, we have to decide whether it is produced at all. We can illustrate the issues involved if we consider only two uses – say $k=1$ and $k=2$. There are then essentially three cases, which are shown in Figure A2.6.

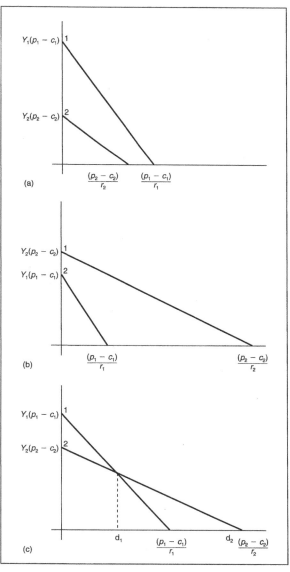

Figure A2.6   Rents from two alternative land uses

In case (a), the rent arising from use 2 will never exceed that from use 1; this also happens in case (b). In case (c), the rent generated by use 1 dominates up to $d = d_1$ and then there is a positive rent for use 2 up to $d = d_2$. How do we distinguish more formally the differences between the three cases? Case (a) and case (c) each have use 1 with the highest rent at the origin *and*, by definition, the steepest gradient. But in case (a), the limit of production of use 2 is reached before that of use 1, and so it is never produced. In case (b), although use 1, by definition, has the steepest gradient, its limit of production is reached before that of use 2, and use 2 dominates because of its advantageous starting rent at $d = 0$. It is clear from the three components of the figure that two conditions have to be satisfied for both goods to be produced:

$$Y_1(p_1 - c_1) > Y_2(p_2 - c_2) \qquad (A2.10)$$

and

$$(p_2 - c_2)/r_2 > (p_1 - c_1)/r_1 \qquad (A2.11)$$

We can then introduce the second notational convention used by Stevens: if a use is not generated, then it is eliminated from the list, and the remaining uses are renumbered again, still consecutively and satisfying equation (A2.9). The geometrical relationships satisfied by these uses are shown in Figure A2.7, and this implies that the following relationships are satisfied:

$$Y_k(p_k - c_k) > Y_{k+1}(p_{k+1} - c_{k+1}),$$
$$k = 1, 2 \ldots \qquad (A2.12)$$

It would be consistent with the notation introduced earlier that the intercepts in the second relation, (A2.11), should be successively labelled

$$d_1, d_2, d_3, \ldots \qquad (A2.13)$$

The two sets of relationships are, of course, obvious extensions of the two-use case given by equations (A2.10) and (A2.11).

Figure A2.7 also shows the distances at which the various rent lines cross. These have been labelled $d_1, d_2, d_3, \ldots$ and it can easily be seen that the land is in use 1 for distances from the market up to d_1; then use 2 takes over up to d_2; and so on. This is von Thunen's main result, and produces the familiar concentric circles of agricultural land use shown in Figure A2.8. The particular example used by von Thunen himself is presented as Figure A2.9.

It is clear from the key to von Thunen's figure, and can be deduced from the original text, that von Thunen is operating at a fairly coarse sectoral level of resolution, but also one which combines his theory of intensities at such a scale with that of land use. A number of his zones involve cropping schemes of different kinds. This argument could in fact be taken further, especially in cases where the input levels (and therefore the intensity levels) could change continuously. The rent functions might then take the form shown in Figure A2.10.

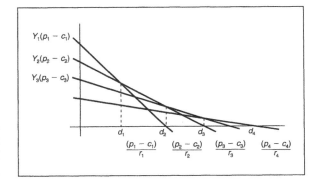

Figure A2.7 Rents from alternative uses, showing limits of use

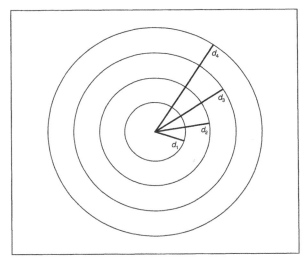

Figure A2.8 Von Thunen's rings for different land uses

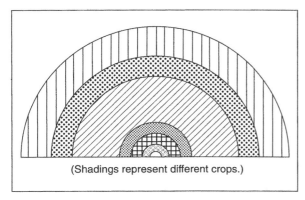

(Shadings represent different crops.)

Figure A2.9 Example of a ring system

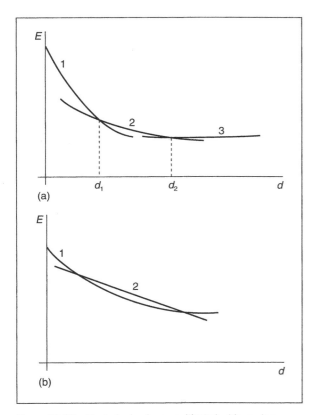

Figure A2.10 Rents for land uses cultivated with varying intensities

In case (a), three economic rents are shown as curves rather than lines, although the essential concentric ring pattern is changed – but with decreasing intensity of use in each ring. Case (b)

shows the intriguing possibility (by using a curve for use 1 and a line for use 2) that a use may disappear and then come back again: use 1 dominates up to $d = d_1$, then use 2 up to d_2 and then d_1 again thereafter.

There is a third problem tackled by von Thunen which involves applying the same methods with a shift of scale: to the individual farm, or possibly to a village. There are essentially three micro-scale cases here, two genuinely so, one really belonging to another category. The 'village' can be considered to take one of two forms: first, it can represent a small 'centre' which is, say, a source of labour for farms that are distributed around it, and which has to be served itself with farm produce; secondly, it could be considered in some circumstances to consist of a group of farmsteads in one location. We treat these cases differently: the first is essentially a generalisation of the von Thunen problem where you have a subsidiary set of market centres as well as the main one, and we consider this below. The second case can be treated as though the village was essentially a large 'farm'. This makes it equivalent to the other genuine micro case which we now consider: the farm itself.

Von Thunen in this case focuses on the effect of distance in another sense. Given that labour and other inputs can be considered to be supplied from the farmhouse, it becomes correspondingly more expensive to provide a particular level of inputs to more distance fields. Using exactly the same argument as at the meso scale, it follows that the farm will be surrounded by concentric rings of decreasing intensity of use. The argument could also lead to different land-use rings if we refine the basic von Thunen argument, as we will see shortly.

RELAXATION OF THE INITIAL VON THUNEN ASSUMPTIONS

One of the beauties of the von Thunen formulation is that, given the basic formula for rent, (A2.1), it is possible systematically to consider relaxing the restrictive assumptions that were made at the outset. Von Thunen himself, with diagrams, showed how to do this for some terms. Figure A2.11 shows the effect of modifying transport cost, in this case through the introduction of a navigable river into the system. The figure shows the distortions from the ideal which result. He also considered the effect of varying grain price (that is, one of the ps) and yields (one of the Ys) with the results shown in the two halves of Figure A2.12.

In a similar manner, it would be possible to consider the variation of production costs, and varying wage levels in different parts of the region, perhaps declining with distance from the centre (a suggestion made, for example, by Chisholm, 1962).

Another feature of Figure A2.11 is the introduction of a small town which von Thunen considers as a 'independent state' in its own right, and this indicates the broad principles for incorporating villages or lower-order settlements into the system.

TOWARDS A MORE GENERAL MODEL

Further extensions take us beyond the simple formula for 'rent' developed by von Thunen. We consider some of the possibilities briefly here, and indicate the outline of a method of dealing with them. They are then taken up in more detail in Chapter 7 in the context of a broader treatment of location theory.

Von Thunen has focused on the effects of transport costs on the outputs of the agricultural system. We have noted in one context – at the farm scale – another feature which involves transport costs: the supply of labour, and it is also possible to add the transport costs of all other inputs. Von Thunen considers some of these, such as the import of manure from urban areas. Cox (1979) considers that these three transport-oriented elements could be considered together. This then ties up with a notion of Chisholm's that the analysis could be inverted. In effect, he argues, von Thunen takes the settlement pattern as given, and then determines the pattern of agricultural land use. In some cases at least, the variability of soils and climate and other environmental conditions may be such that the land use is largely determined by these and what is at issue is the location of farms and settlements. The location of a village, say, can then be determined in relation to its resource inputs: agricultural products, water, fuel, building materials and so on, and the analysis can then proceed in the manner of Weber and other location theorists of that school. This is a fascinating point. It anticipates the next section and should be borne in mind in that context. It also illustrates the general point that most partial location problems are determined by choosing

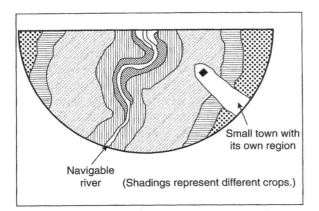

Figure A2.11   Modified von Thunen rings: transport costs

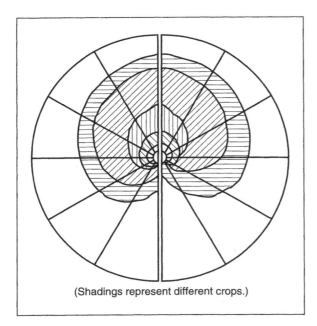

Figure A2.12   Rings distorted by varying grain price

something to be fixed. In the end we need a comprehensive general equilibrium analysis.

One of the most serious criticisms that can be made of the von Thunen system, and which opens up a general type of question in location theory, is his assumption of a fixed price and, more importantly, inelastic demand: that any amount of product can be taken into the market. There is an obvious need to generalise this notion: to have demand curves at the market point, and supply curves constructed out of the production possibilities of the agricultural system. Then, quantities, land use, intensities and prices could be determined within a model. It is also necessary to address the question of trade: it may not be an optimum use of land to produce everything needed for a particular market point. This market needs to be linked with others, as is the case in reality, through trade relations so that in its own region, it is possible to concentrate on land uses for which the region has a comparative advantage. This is obviously a function of transport costs between regions as well as other more obvious locational features.

In general, we are seeking an approach to the agricultural land-use problem within which many of the (assumed) 'constants' can vary, and the restrictions on numbers of settlements and market points can be removed. It would then be possible to integrate the different scales considered by von Thunen: the farm would be the lowest order of market point – for part of its own product, the village the next, and so on. When the system is extended to include a number of settlements, the trade flows between settlements within the region of the main market town should also be considered. This may allow for more specialisation in particular areas than is permitted in the von Thunen system. This problem has been tackled in a number of ways. Later in the appendix, we consider some of the classical approaches to market area analysis which reaches its fullest fruition, in some sense at least, in central place theory. But it also needs to be integrated with the Weberian analysis which we consider below. So, our consideration of some of these more general issues must be postponed. However, it is worth giving one clue on how some of these problems can be resolved, and we do this briefly in the next section.

1	2	3	4	5
6	7	8	9	10
11	12	13	14	15
16	17	18	19	20
21	22	23	24	25

(Zones are numbered, j = 1, 2, 3 ...)

Figure A2.13 An empty grid for a zone-based von Thunen analysis

A change of spatial representation: the basis for a general model

Consider a region divided into zones, as shown in Figure A2.13. Label zones by $j = 1, 2, 3, \ldots$ We can then consider the economic rent arising in zone j, per unit area, for land use k:

$$E_{jk} = Y_k(P_k - c_k - r_k d_j) \tag{A2.14}$$

where d_j is now the distance from zone j to the market point. It is straightforward now, to expand the terms in this equation to include transport links, for products, labour or inputs, with a number of centres (or indeed one, if it is desired, to reproduce the original von Thunen results in this representation).

Given a single market centre, concentric rings would be generated as shown in Figure A2.14. It is now easy to visualise how much more complicated patterns can be constructed, and without any essential difficulty. It is possible to use mathematical programming to solve a much more general class of problems while still retaining a number of von Thunen's basic ideas. This idea is pursued in Chapter 7.

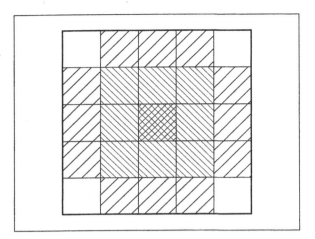

Figure A2.14 von Thunen rings in a grid

The location of industry: Weber

INTRODUCTION

Weber's classic text on the location of industry was published in German in 1909 and in English translation in 1929. It is a masterpiece of exposition and rightly has a major place in the history of location theory. In von Thunen's case, it is interesting to read the original book, but most helpful to read the digests of commentators ahead of such an exercise. In Weber's case, his book belies its date of writing and can be read with pleasure and the gain of much insight.

Weber was another modeller. He postulated an idealised world and then, in a similar manner to von Thunen, explored the consequences of systematically relaxing his assumptions. In the next section we present his basic problem, the results of which have a wider variety of applications than Weber would have anticipated, given the different kinds of organisations in a modern economy compared to his own day. This forms the basis for the construction of various indices which characterise the location behaviour of firms in different kinds of industries. Two extensions, one to deal with spatial variations in labour costs, the other with agglomeration economies, are reported, the task of relaxing the assumptions is explored, and some concluding comments are made.

THE BASIC WEBERIAN PROBLEM

Weber's basic model for the working of an industrial firm was as follows. It took in raw materials that were location-specific (which may be half-completed products, but this more complicated case is reserved for later discussion). These materials were called *localised*. The firm might also use ubiquitous materials, such as water or clay. All these materials were transformed (using capital and labour) into a product that was sold at a particular consumption point. The locations of the different raw materials were also known. It was assumed that the cost of labour, capital and ubiquitous materials were not location-dependent and therefore did not enter into the location decision of the firm. The assumption about labour is relaxed below.

The basic situation facing the firm, assuming for the sake of illustration that it uses raw materials from two different locations, neither of which coincides with the consumption point, is shown in Figure A2.15: M_1 and M_2 are the sources of raw materials and C is the consumption point. Where should the firm locate? Consider the trial location at the point P in Figure A2.16. Weber argued that the only spatially variable costs were transport costs, and therefore the firm would locate so as to

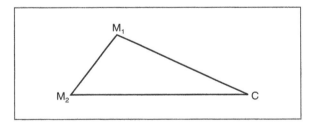

Figure A2.15 The basic Weberian problem

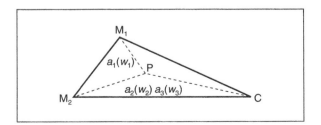

Figure A2.16 Optimum location of the factory

minimise these. He assumed that transport costs would be proportional to weight and distance.

In Figure A2.16, the distances from P to M_1, M_2 and C are shown as a_1, a_2 and a_3 respectively and w_1 and w_2 are the weights of raw materials needed to produce weight w_3 or the product. Total transport costs are then given by

$$Z = a_1 w_1 + a_2 w_2 + a_3 w_3 \qquad (A2.15)$$

and this is to be minimised. Note that the co-ordinates of the point P can vary continuously over the space which has been defined.

It turns out that there is a mechanical solution to this problem, and a geometrical one in the case where there are only three 'pulling' points involved in all. There is no analytical solution to this as a mathematical problem though the advent of modern computers has led to an iterative algorithm being invented to solve the problem. Here, we restrict ourselves to the two solutions considered by Weber.

The mechanical solution is pictured in Figure A2.17. Imagine weights located at the ends of strings running over pulleys at each vertex of the triangle. The weights at each vertex are proportional to the total transport costs involving that vertex as shown. If the three strings are joined in the point P, then the equilibrium location of P provides the solution to Weber's problem and minimises total transport costs. (The apparatus used by Weber was apparently originally devised by Varignon to demonstrate the parallelogram of forces in classical mechanics.)

The basis of the geometrical solution, much displayed in textbooks, is shown in Figure A2.18. Construct a triangle whose sides have the lengths

of the total transport costs involved and let the angles between the sides be labelled α, β and γ as shown. Then it can be proved that the angles around the point P, shown in the second part of the figure, are $180 - \alpha$, $180 - \beta$ and $180 - \gamma$. A construction can then be given to locate the point P using compasses. However, this derivation is only of historical interest for two reasons. First, it has been overtaken by the computer algorithm mentioned earlier. Secondly, it does not generalise to the case where more than two sources of raw materials are involved. Such a case is shown in Figure A2.19. The mechanical device does still

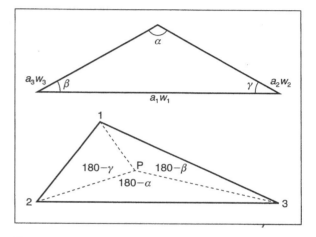

Figure A2.18   The geometrical solution

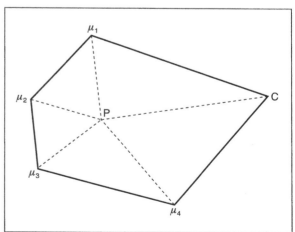

Figure A2.19   More than two sources of materials

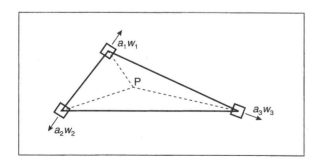

Figure A2.17   The Varignon frame solution

generate the optimum solution, and the computer algorithm is easily extended also. Too much significance, therefore, should not be attached to what is often called the 'Weberian triangle' as such: the more general case is more important.

The next question to tackle is: how can we characterise different kinds of firms using the concepts now introduced?

THE MATERIALS INDEX AND LOCATIONAL WEIGHTS

We begin by examining in more detail the nature of production processes in relation to the various weights involved. Weber distinguishes *gross* raw materials from *pure* ones. In the former case, there is at least some weight lost during the production process. In the case of materials which are used as fuel, for example, like coal, all the weight is lost relative to the final product. In the case of a pure raw material (such as diamonds) all the weight forms part of the weight of the product. The next part of the argument is to recall the role of ubiquitous raw materials which are assumed to be available at the site. These may be used in such a way as to add weight to the product (local timber, water, fixing of nitrogen, and so on). Thus, the weight of the product may not bear any direct resemblance to the total of the weights of the incoming *localised* raw materials.

Let w^1, w^2, w^3, ..., w^n be the weights of n raw materials needed to generate a weight w^* of the product. The total weight of the raw materials is

$$W = \Sigma_m w^m \qquad (A2.16)$$

and Weber defined a *materials' index* for an industry as

$$M = W/w^* \qquad (A2.17)$$

This can range from zero, if no raw materials are used at all, to very high figures if the process involves a large volume of weight-losing raw materials for a relatively small weight of product.

Weber notes that the weight used of ubiquitous materials is only significant in so far as they contribute to the weight of the final product. He defines therefore the total weight-generating transport costs as

$$W' = \Sigma_m w^m + w^* \qquad (A2.18)$$

and the location weight of the industry as

$$L = W'/w^* \qquad (A2.19)$$

Simple algebraic manipulation shows that

$$L = M + 1 \qquad (A2.20)$$

and hence only one of these needs to be used.

To save confusion for the reader who is tackling Weber (and some commentators) directly, note that these indices are sometimes defined in relation to a unit weight of the product. This simply involves setting w^* to 1 in the above equations and defining w^k as the amount of the raw material used to produce a unit weight of the product.

How can we now use the materials index to characterise the locational behaviour of different kinds of firms? Weber begins by making some general observations and then considers a series of cases that cover all possibilities. When the materials index is high, the firm is attracted towards the sources of raw materials. When the material index is less than one, then the location must be at C, the place of consumption. If only pure materials are used, the materials index is less than or equal to one, and so the optimum location will tend towards the consumption point. It is the increasing proportion of weight-losing materials that can pull P towards the sources of the materials.

Weber considered, in more detail, three kinds of case:

(i) Ubiquities only. In this case, P is always at C.
(ii) Localised pure materials, with or without ubiquities. P is always at C except in one special and interesting case. The exception occurs when there is only one localised pure material and no ubiquities. The materials index is then one, and a little thought shows that anywhere along the line joining the materials source and the consumption point is an optimum solution. This is an example of a case where there is no unique solution to a locational optimisation problem, and we will have occasion to examine such situations in another context in Chapter 7.
(iii) Use of weight-losing materials, with or without pure materials or ubiquities. There are a number of interesting distinct cases here:

(a) With one weight-losing material alone, the optimum location is at the material source.

(b) If ubiquities are added to this case, then the location remains at the materials source until the volume of ubiquities increases to the point where the materials index becomes less than one.

(c) When there are several weight-losing materials, the optimum location will be somewhere in the polygon formed by the material sources and the consumption point, and the methods described in the last section can be applied.

THE INTRODUCTION OF VARIABLE LABOUR COSTS

The next step in Weber's argument is to consider relaxing some of the assumptions associated directly with the last two sections. However, it is appropriate to examine first the new tools he makes available to relax two of his major assumptions, concerned with labour and agglomeration. We do this in the next two sections and then return to the issue of assumptions in general.

We should begin by remarking that Weber's treatment of labour reveals an apparent inconsistency in his spatial representation. It has always been argued (by him) that labour is available at a number of fixed locations. There is therefore an implied assumption that the cost to the labourer of transport to the workplace is borne by the worker and has no effect on wage rates or labour availability. (Or, as implied in another context by Weber later, that labour relocates to the optimum location of the firm.) This seems unlikely to be true in practice, and there seems no reason in principle why labour should not be treated simply as another input. This is an issue to which we shall return. Meanwhile, we accept Weber's form of assumptions. When he tries to introduce variable labour costs, he does this by assuming that different wage rates are payable at the fixed points already referred to. The analysis then proceeds as follows.

Suppose that the optimum location of the firm has been found by the procedure described in the section on the basic Weberian problem. It would then be possible to construct *isodopanes*, which are contours of equal additional total transport cost. That is, if the location of the firm was shifted, the additional transport cost could be calculated. If this is done for large numbers of points at some standard intervals of cost (which is a difficult thing to do in practice, but is straightforward conceptually), then the isodopanes can be drawn. An example is shown in Figure A2.20. Let L_1, L_2 and L_3 be the locations of labour supply. Let there be a definite maximum amount in labour costs that can be saved, potentially, by switching to a new location at one of these points. This amount defines the *critical isodopane*, and this has been drawn as a dashed curve in Figure A2.20. If there is a labour supply point inside this curve, then it will pay the firm to shift to that point if the saving in labour costs exceeds the increase in transport costs. This shows in general how such shifts might occur.

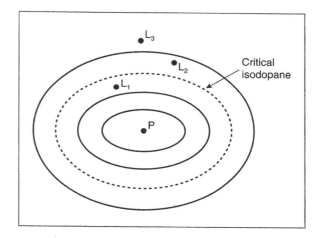

Figure A2.20 Isodopanes: contours of equal transport costs

AGGLOMERATION ECONOMIES

Weber recognises the fact that he has a very partial theory in that he is working with an individual firm largely considered to be independent of all other firms. His second major extension, therefore, is to consider tendencies to agglomeration. He identifies a number of factors which may make it attractive for plants to group together – either plants within a particular industry, or plants within different industries. The first is the possible, indeed likely, existence of scale economies. As plant size increases, there are likely to be scale economies arising from better division of labour, larger machines, more effectively shared overheads (which Weber considers under a separate heading) and so on. The remaining factors relate more to the existence in the locality of firms from other industries. These include access to different kinds of equipment, some of which may be manufactured or developed by firms in other industries; division of labour for all the firms in the locality – the possibility of a wider range of skills being available for example; and improved marketing possibilities.

He also notes that tendencies to deglomeration can be dealt with under the same heading, but that in this case there is no need to distinguish different factors because their effects will be grouped under the increased rent for land which firms have to pay. This arises from the competition for land from firms, in different industries, which is likely to bid up its value.

When the results of these factors are combined, their benefits, if any, can be estimated, for particular firms. They can then be treated in the same way as variations in labour costs. The isodopanes representing contours of increments in transport costs away from the optimum transport-oriented location can again be drawn. This time, we can focus on the critical isodopanes of a number of different firms. If they have areas of intersection, as shown in the case in Figure A2.21 for example, then all the firms could move to a point within the area of intersection and take advantage of the agglomeration economies. The possible location points are represented by the shaded area in the figure.

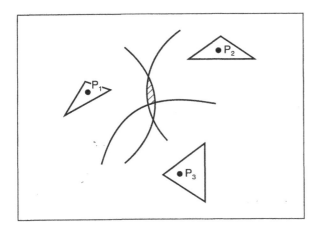

Figure A2.21 Isodopanes and agglomeration economies

Weber then goes on to discuss a number of refinements. This analysis, for example, can be combined with that of shifts towards cheaper labour. It is also possible that once a firm has shifted for either of these reasons, then it may also be profitable to shift from one particular source of materials to another. There are obviously particularly high benefits to be gained if points of agglomeration coincide with cheaper labour points; and further gains still if there are alternative materials supply points which modify the costs involved in moving away from the minimum transport-cost point.

Weber notes that firms for whom agglomeration economies are more likely to be substantial are those which have a high value-added. These are likely to be labour- or capital-oriented rather than being highly dependent on the import of raw materials.

He concludes the discussion on agglomeration by noting the tendencies in development that were visible to him at the time of writing. These sound familiar, at least up to the fairly recent past: that population densities were increasing and that transport costs were decreasing. Both of these phenomena encouraged agglomeration on an increasing scale.

RELAXATION OF ASSUMPTIONS: TOWARDS A GENERAL SYSTEMS APPROACH

Weber considers at an early stage the possibility of relaxing the kind of 'uniform plain' assumptions he has made with respect to variables such as transport costs. This can be done in a relatively straightforward way. In effect, his pictures, as were von Thunen's, are topologically transformed in a way that is familiar to us with modern techniques of analysis.

He also considers a number of relaxations of the assumptions which determine the nature of his firms. For example, he recognises that some firms manufacture products which are directly the inputs of other firms. Or that firms exist which input raw materials and manufacture from these two or more different intermediate products which then form the inputs to other firms. These situations can obviously be analysed with the same kinds of techniques as presented for the simpler case, and he presents a number of case studies.

His biggest attack on the relaxation of assumptions, however, comes in the last two chapters where he considers what he calls 'the total orientation' and then a final examination of the nature of manufacturing industry in the economic system as a whole. The argument is very impressive. In effect, he adopts the approach of systems analysis and considers all possible interdependencies. He anticipates the development of much theory beyond the confines of industrial location analysis and we recognise the germs of many ideas which emerge more fully from other authors in later chapters. It is worthwhile attempting to give the flavour of this, though the reader is very much encouraged to read the original.

He begins with a discussion of the organisation of a particular industry, noting that it is often made up of technical processes that are independent though obviously, in the end, linked. This provides a case study of the type referred to in the previous paragraph. What is more interesting, from a modern viewpoint, is when he focuses on the interdependence of different industries. He recognises the possibilities of multi-product industries; that different industries may be dependent on the same set of materials, and particular intermediate products; and that the basic product of one industry may be an input to another. (He cites the example of 'wrapping material'.) This third point he describes as a *market* connection of one industry with another. One firm, in effect, becomes another's consumption point. In this case, the firm manufacturing the auxiliary product can obviously be drawn towards the location of other industries. There is the beginning, in his analysis, of the notion of an input–output model.

The final chapter is particularly interesting in the way many of the current problems of location theory are anticipated. Weber notes that he has made a number of assumptions about the location of parts of the whole economic system not concerned with manufacturing but which determine its location: (i) places of consumption; (ii) the location of material deposits; (iii) the location of labour supply; and (iv) that this labour supply is available in unlimited quantity at constant cost. He sets about exploring the interdependence of all these elements.

The recognition of interdependence, and the way it is expressed, could well have appeared in a book written 50 or 60 years later than it was. He writes:

> There appears at once before our imagination the picture of a circle of force which it seems hardly possible to break through. The location of the places of consumption, the labour locations, and the material deposits, which supposedly determine the location of industries, are themselves resultants of this very same industrial orientation. For each particle of industrial production which moves to a certain place under the influence of locational factors creates a new distribution of consumption on account of the labour which it employs at its new location, and this may become the basis of further locational regrouping.

He argues that in order to break through this circle, it is necessary to assume that there is an equilibrium at a given time and to consider the modifications which can take place from that position. He produces, in effect, the framework for the dynamic analysis, albeit on comparative static terms, of a comprehensive model.

He then takes this argument further by considering that the locational structure of the economy is determined, partly for reasons of historical development, on a hierarchical basis. This begins with the location of agricultural production which can be analysed, he recognises, using von Thunen's theory. This also generates a population pattern which forms consumption points for industry also. The next sector to develop is what he calls primary industry (*not* primary in the sense of Chapter 2) which serves the agricultural sector. Then there is a secondary industry which serves this primary sector. He identifies other groups concerned with distribution of goods and with administration, and that these two population groups generate a further industrial sector to meet their needs. He recognises the multiplier effects of these developments in turn and has the basis of a sophisticated multi-sector economic-base model or a rudimentary input–output model. He even recognises that the high complexity of the 'system' is such that his model of development through a sequence of strata or hierarchies is likely to be inadequate. He would probably have been astonished at what he could achieve in tackling some of the questions he raises if he had been able to use the tools that can now be used by the theorist in human geography or urban and regional economics. But we should recognise the magnitude of his achievement, beyond his study of industrial location as such, in identifying the nature of the location problem as a whole.

CONCLUDING COMMENTS: SPATIAL REPRESENTATIONS AGAIN

We have seen that, implicitly, there is some muddle in Weber's treatment of space. Some elements are located at fixed points, while his firms can locate at optimum variable points. There will, of course, be an increasing number of 'fixed' points in his system as agglomerations develop. We can see in our explicit algebraic treatment of the Weber problem in Chapter 7 that the assumption of variation in continuous space for the location of firms creates difficulties for a generalised theory. Once again, there is no inherent reason why the spatial basis should not be a discrete zone system, with centroids forming a lattice of points, which then lends itself to analysis through more powerful mathematical tools, such as mathematical programming.

If there is a main criticism to be levelled it is perhaps that Weber does not face up to the idea of firms within an industry competing with each other for their markets, and the locational consequences of this competition. This is a separate topic which is also a useful preliminary to a discussion of central place theory.

In concluding in the previous section, we have recognised the power of Weber's general approach to location problems. It is perhaps appropriate to conclude by returning to the importance of the specific problem he formulated: the optimum location of the firm relative to a given set of material sources and a consumption point. If the discrete spatial representation is used, it becomes relatively straightforward to extend the problem to include several consumption points, and there seems to be no reason in principle why all the inputs of the firm, including labour, should then not be treated on a similar basis. This makes it possible to incorporate other kinds of inputs

without resorting to the transport isodopanes as a computational device. It is then possible to see Weber's firm in a very general way, not simply as an organisation within manufacturing industry. We can see in Chapter 7 that perhaps the most important modern applications of his ideas, ironically in some ways, involve finding the optimum location of public facilities relative to a large number of consumption points.

Weber solved some major problems. His influence has generated many other ideas. The only surprise is that, given the existence of his book in 1909, some of these were not developed much earlier.

Market areas and competing firms: Palander, Hoover and Hotelling

INTRODUCTION

Neither von Thunen nor Weber, at least in the formal parts of their theories, give any consideration to competing organisations with respect to their locational behaviour. It is therefore appropriate to consider some early contributions to this question. This is useful for its own sake: some new insights are obtained; and it is also a useful preliminary to the account of central place theory in the next section.

We restrict ourselves here to the contributions of three authors, taking them almost, but not quite, in chronological order: Palander's main work was published in German in 1935; Hoover's in 1937; and Hotelling's famous paper on stability under competition in 1929.

Weber assumed that all the goods produced by his single firm could be sold at one consumption point. To consider the question of competition involves a shift of spatial representation: that there are a large number of consumers; and they can be considered either to be distributed continuously in space or across a large number of fixed points (say on a lattice, or at the centroids of a discrete zoning system). A firm might then have a number of 'good' locations in relation to material or other inputs in a Weberian sense. But how does it compete with other firms? This question is usually addressed in relation to the number of consumers it serves, spatially, and this region is known as the market area of that firm. This notion, of course, assumes that consumers purchase their goods from the firm which supplies them most cheaply. We will see in other contexts later in this appendix and in the main chapters that this in turn is too simple an assumption, but it will suffice for now. We consider examples of the three authors' ideas in turn.

PALANDER

Palander considered two firms offering the same good to a spatially distributed population of consumers. He discussed a number of different cases which are distinguished by the prices and transport costs associated with the good from the two sources. All that is necessary for present purposes is to report some of the main results. The methods of construction used by Palander are of less importance for later.

Let the two firms be located at P_1 and P_2 as shown in Figure A2.22a. Let the price of the goods at the point of production be p_1 and p_2 respectively, and let r_1 and r_2 be the transport rates from the two firms. (These may be different because the two firms have different pricing policies in relation to delivery for example.) Palander analyses the form of the boundary which distinguishes the two market areas in a number of cases, three of which are presented in Figure A2.22.

In case (a), both basic prices and transport rates are equal, and the market boundary is a straight line bisecting the line joining the firms and perpendicular to it. In case (b), the basic price at firm 1 is less than that at firm 2, but the transport rates are the same. This results in a hyperbolic boundary as shown in A2.22b. In case (c), the basic prices are again the same, but the transport rate is greater at 1 than at 2. In this case, the boundary is a circle around the location of firm 1 – somewhat off-centre. This means that, because of the cheaper transport costs from firm 2, there comes a point where its market area extends beyond the circle at the other side of firm 1.

The main result of the set is the first one, which will be used throughout the discussion of central place theory below.

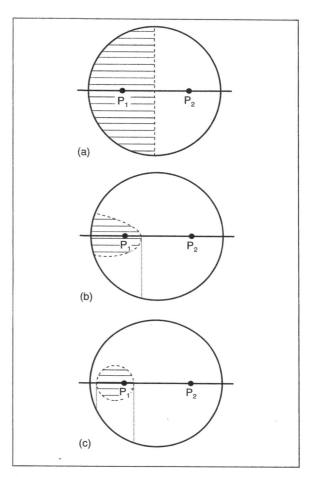

Figure A2.22 Palander's market areas. (From Birkin and Wilson, 1986b)

HOOVER: MARKET AREAS FOR MORE THAN TWO COMPETING FIRMS

Hoover provides a useful example of the derivation of market areas graphically for three competing firms, with a method of analysis which could obviously be extended. He again considers the possibility of different production costs for firms at different locations, and the analysis would easily extend to differing transport rates. Each of his firms are considered to be optimally located in a Weberian sense with respect to their inputs (and it is in this sense that their production costs may differ: their materials' sources may be differing distances away, for example, in each case; or labour costs may differ). Hoover constructs

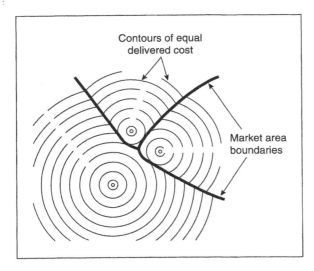

Figure A2.23   Hoover's market areas

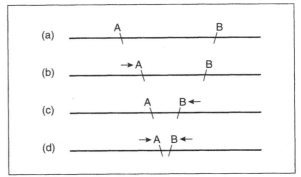

Figure A2.24   Hotelling/Alonso: two ice-cream men on a linear beach

contour lines of equal delivered cost, and this leads to diagrams such as that shown here as Figure A2.23.

It is easy to see that market area boundaries can be deduced from the contours, and these are shown as heavy lines for the three firms in Figure A2.23.

HOTELLING: STABILITY UNDER COMPETITION

Hotelling's example is now probably so well known as barely to need description. It is usually presented in the form developed by Alonso (1960). He considers two ice-cream men to be located on a linear beach. It is considered that the demand for ice-cream is inelastic and that people are distributed uniformly along the beach: each person consumes the same amount of ice-cream per day.

Consider Figure A2.24. There is an intuitive optimum solution to the location problem: they 'should be' at the quartiles of the beach as shown in Figure A2.24a. Then they each have half the market and it is easy to see that consumers, on average, travel the minimum distance for their ice-cream.

But is this situation stable under competition? If A moves towards the centre, as in A2.24b, then he captures more than half the market. The only way in which B can respond is by moving towards the centre also (A2.24c). It is then easy to see that the stable competitive equilibrium occurs when

both ice-cream men are at the centre of the beach (A2.24d). They each have half the market again; the situation is stable; but the consumer on average now travels much further for ice-cream!

Hotelling's argument has been suggested as the basis of a new contribution to the theory of agglomeration: it is not so much agglomeration economies that brings it about in such cases, but the outcome of the competitive process. The gathering together of shops of certain types is often cited as an example.

The situation becomes less simple if demand is not inelastic. If the case is more like that treated by Palander, with a basic price and a transport cost incurred by the consumer, and if demand falls as the total cost increases at certain points along the beach, then it can be shown that the position is reached where the stable equilibrium is at the quartiles.

This simple example illustrates a point that has been raised before: the importance of building in more realistic demand relationships. There may, however, be circumstances where the Hotelling agglomeration effect does operate.

Central place theory

INTRODUCTION: THE BASIC ASSUMPTION

Central place theory is associated with the names of two Germans: Christaller, a geographer, and Losch, an economist. Their systems have much geometry in common and although we will start with Christaller's theory, we will use some of the clarifications of this theory articulated by Losch and other authors. Spelling out the geometry in this way enables us to offer a relatively brief account of Losch's work later.

There is much controversy about the assumptions of central place theory. Initially, we will take as simple as possible a view of these in order to present the main ideas. A more critical review will be offered later. There are essentially two kinds of assumptions: those concerned with the behaviour of the various agents in the system, broadly speaking, of firms and consumers; and those related to the backcloth on which the action takes place.

Firms are assumed to maximise profits in producing goods and services and selling them to the population. People are assumed to minimise transport costs in purchasing the goods and services they require. The geographical system is assumed to consist of a uniform distribution of population and a ubiquitous transport system which is such that travel is equally easy in all directions. The population is considered uniform both in density and purchasing power. Firms are assumed to locate in centres (which are treated as points). These centres are settlements which also each have a population (and hence, as soon as centres appear, the uniform *density* assumption is in a sense modified).

In Christaller's original formulation, the uniform population was essentially the rural population based on farms and agricultural work. The settlements are villages, towns and cities. Clearly the balance of size, influence and importance of urban and rural populations has changed since Christaller carried out his research on southern Germany in the early 1930s. There are new features of the spatial economy, such as firms becoming more interdependent and supplying goods to each other rather than simply serving a largely rural community. However, settlement populations, and their demands, can be taken as a proxy for these kinds of effects. This shows that the assumptions can often be given new interpretations and that they should not necessarily be considered very restrictive. Indeed, it turns out, as we will see in our critical assessment later, that it is other features of the model which

are more damaging, and so a long discussion at this stage of what are usually taken as the assumptions is not likely to be very productive.

BASIC CONCEPTS

The underlying economic assumptions are in essence very simple: that for any good, there is a demand curve relating price and quantity purchased as shown in Figure A2.25. It is assumed that each consumer pays the price for the good and the transport costs so that the quantity purchased is reduced for consumers at greater distances from a centre. Indeed, there will be a distance from a centre beyond which the amount purchased falls to zero (Figure A2.26). This is what Christaller calls the *ideal range* of the good. In practice, it is likely that, for consumers on the edge of the market area of a centre, an alternative and nearer centre is going to be available. This competition from other centres reduces the ideal range to the *real range*.

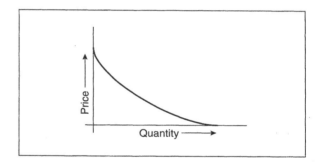

Figure A2.25 Demand curve for a good

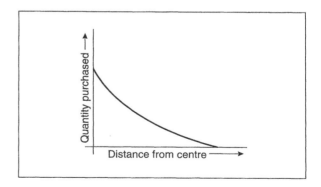

Figure A2.26 Distance from a centre at which demand falls to zero

We will explore the details of this in the next section.

The other concept which is important at the outset is that of *threshold*: this is the minimum area (or the radius of the equivalent circular area, or its population – whatever is taken as the convenient measure) that will support the provision of that good or service. This, in effect, is an assumption about the form of the production function: there are some basic (fixed) costs to be incurred, there may be scale economies which generate a minimum size, and the good or service can only be supplied at a reasonable unit cost if production is on a certain scale.

There is considerable discussion in the literature about the possibility of excess profits. In principle, if the real range for a good exceeds the threshold, then it can be argued that excess profits can be made. However, some authors (e.g. King and Golledge, 1978) argue that these will be competed away *within the given market area* by the introduction of new firms at the centre and a corresponding lowering of price until the *average* unit cost (including normal profits) equals unit revenue. On the whole, however, this controversy does not affect the general argument.

We can now begin to approach the idea of the order of a good. Generally speaking, goods with a low range, like groceries, are lower order goods, and vice versa. This will be partly determined by the value relative to weight; partly by consumers' preferences. The threshold is likely to vary with the range (though Beavon (1977, p. 35) notes the possibility of a number of alternative functional relationships, while King and Golledge argue that the threshold always equals the range if it is assumed that there are no excess profits). It is then possible to rank goods according to their range, from highest to lowest. There is no reason why this should be an evenly stepped distribution; nor is it easy to say *a priori* what the main sources of revenue are – rank (and range) does not necessarily correlate with total revenue attracted. Very specialist goods or services, for example, may generate a relatively small total revenue but have a large range (and threshold).

The final step in this broad argument, therefore, is to observe that, corresponding to the orders of goods and services will be orders of centres. The next step is to see how the location and spacing of these can be arranged.

THE BASIC GEOMETRY OF CENTRAL PLACE SYSTEMS

On the uniform plain assumptions, it follows that the pattern of the spatial arrangement of centres will have some kind of symmetry. If we begin with the assumption that, for one good say, the market area of each centre is circular and just touches adjacent areas, we get the kinds of patterns shown in Figure A2.27. It turns out that only these three patterns are possible bases for central place theory for reasons that will emerge shortly and we will discuss the nature of these patterns in more detail.

In each pattern in Figure A2.27, there are areas of the plain which are not served by centres. Christaller made it an assumption of his theory that all areas should be served. (It is not clear that this is always consistent with profit-maximising assumptions but let us assume that marginal costs and revenues are such that it is.) This can only be achieved if the circles are increased in radius until they overlap. Examples are shown in Figure A2.28 for the same three cases that were presented in Figure A2.27. The circles intersect in regular polygons: triangles, squares and hexagons respectively. (Note the duality between a and c of Figure A2.27: in a, the centres of the circles are the vertices of hexagons within the equilateral triangle base; in c, vice versa.) These form the possible market areas for central place theory. We briefly explain the geometrical reasons for this before proceeding.

There is a valuable discussion of the principles involved in Haggett *et al.* (1977, pp. 55–57). They define efficiency of different types of polygons for 'space packing' in terms of two dimensions: movement from centre to periphery and length of boundary. A circle in this discussion is considered to be the limiting case of a regular polygon (with an infinite number of sides). Efficiency of movement is measured as the distance to the furthest point (for a given amount of area enclosed). First, it is noted that regular polygons are always more efficient than irregular ones of equivalent numbers of sides. Secondly, regular

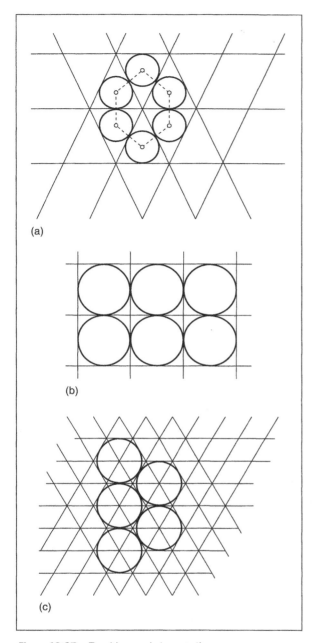

Figure A2.27 Touching market areas: three cases

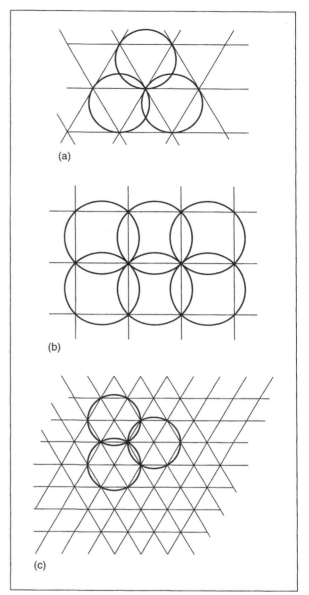

Figure A2.28 Overlapping market areas to cover space

polygons become more efficient as the number of sides increases, with the circle being the most efficient. Thirdly, there are only three regular polygons that can be used to fill a plain with equal-area units: the triangle, the square and the hexagon (and hence the choice of bases in Figure A2.27).

It now follows that the way to cover an area most efficiently with regular polygons from the circular market areas we started with involves expanding them from, as it were, Figure A2.27 to Figure A2.28, and it is Figure A2.28c that results in the hexagonal structure that is most well known. Haggett *et al.* (1977) report that the hexagon is four-fifths as efficient as a circle, and much more

efficient than either squares or triangles. Note that the form of the regular hexagon that is generated turns on the spatial arrangement which has been assumed for the centres as noted earlier. Recall that the centres which generate the hexagonal areas are the vertices of a lattice of equilateral triangles, and this is a most important result. (The squares are generated by a square lattice and the triangles, ironically perhaps, by vertices of a hexagonal lattice.) Henceforth, we concentrate only on the system with hexagonal market areas whose centres are located on a lattice of equilateral triangles. Note now that the real range is not the same in all parts of the perimeter of the market areas.

The main geometrical measurement property to note at this stage is that which relates the maximum real range, say a (the radius of the circle after it had been expanded, i.e. the distances from the centre to a corner of the hexagon) to the spacing of the centres, say b. The relationship is exhibited in Figure A2.29 and elementary trigonometry shows it to be

$$b = a.\sqrt{3} \qquad (A2.21)$$

or conversely, of course,

$$a = b/\sqrt{3} \qquad (A2.22)$$

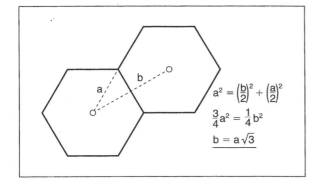

Figure A2.29 Real range and centre spacing

HIERARCHICAL CENTRAL PLACE SYSTEMS

Christaller's marketing principle: the k = 3 system

The argument of the last section leads to a covering of the region with a set of hexagons (as shown in Figure A2.30), which are the market areas of a set

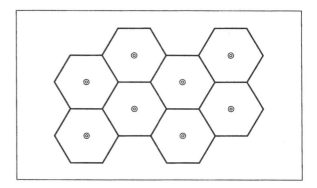

Figure A2.30 Space covering hexagons for a set of centres of the same order

of centres all assumed to be of the same order. What goods and services are produced and traded in these centres? There are two cases of immediate interest. First, if the threshold of a good is greater than the maximum real range of centres in the system shown (which was defined as a), then the good will not be produced. It needs a higher order centre. Secondly, if the ideal range of a good is less than a, then the good will be produced at the centre (because its threshold, by definition, must be less than or equal to the range if it is to be produced at all), but the whole area will not be covered. Complete coverage can only be attained through the introduction of lower order centres. Thus, we have constructed an argument for a hierarchy of centres to exist, defined in relation to the orders of the goods they offer. The next question, then, is how this hierarchy can be constructed. In this and the following three subsections, we present the basic geometries of alternative hierarchical structures, and then discuss some of their implications.

Given the system in Figure A2.30, and given an acknowledged need for lower order centres, where should they be located? The points of Figure A2.30 which are the least likely to be served by the higher order centres are the vertices of the hexagons which define the market areas. These points are also equidistant from three of the current-order centres. These locations are the first ones we explore.

At this point we adopt an aspect of Christaller's terminology that has also been adopted by most other authors. It is clear that, in defining a

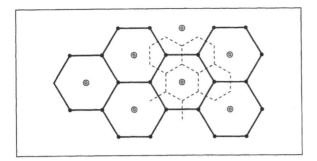

Figure A2.31 Lower order centres added

hierarchy, it is necessary to start somewhere: the first set of centres introduced plays a crucial role in determining the geometry of the whole system as we will emphasise later. Christaller called these B-centres, and the next lower ones, K-centres. Since K-centres are equidistant from B-centres, and since B-centres also trade in lower order goods, then given the assumptions that consumers travel to the nearest centres which supply the goods, the market areas in lower order goods will be shared in the manner indicated in Figure A2.31. That is, we introduce another set of hexagons which represent the market areas – the maximum real ranges – for lower order goods.

Christaller called the notion which generates the picture shown in Figure A2.31 the *marketing principle* because it minimises the number of market areas (or centres) involved in supplying the next lower order of good from the one already defined (and also the distance travelled by consumers). We will see exactly why this is so below. The K-centres can be considered to be dependent on B-centres. This arises because they will have their own populations and they will be served with B-order goods from B-centres. Since each K-centre is served by three equidistant B-centres, each B-centre can be considered to have a one-third share in this trade to K-centres. Thus, since a B-centre also serves itself in lower order goods (and is in this sense also a K-centre) it serves itself and one-third of each of six K-centres. This is the equivalent of three K-centres in all, and this is why this is known as a $k = 3$ system. The same number also applies to market share for the rural population on the uniform plain in respect of K-order goods: Figure A2.31 makes it clear that any

one B-centre serves its own K-hexagon and one-third of that of each of the surrounding K-centres, again making a total of three in all.

The same arithmetic, which can be confirmed by further scrutiny of Figure A2.31, shows that k is also the ratio of the number of K-centres to B-centres. The argument can then be further extended in either direction for other orders of goods. Christaller, in his study of southern Germany, introduced three higher orders than B and three lower orders. For the $k = 3$ systems, we can see from equations (A2.21) and (A2.22) and from Figures A2.30 and A2.31 that if a is the maximum real range (and also the length of a side of the hexagon because of the nature of the equilateral triangle lattice), then it now also measures the spacing between K-centres. Then $a\sqrt{3}$ is the spacing between B-centres, and this procedure can be continued, up or down, using $\sqrt{3}$ as a ratio or factor. This regularity arises from the way in which each market area at successively lower levels bisects the ones at the higher level. The spatial arrangement of the whole system is shown in Figure A2.32.

The transport principle: the $k = 4$ system

As a first alternative principle for locating the subsidiary K-centres, Christaller considered that the transport routes between B-centres might be of paramount importance and that therefore there may be circumstances when the K-centres would be located on lines joining B-centres. They are again located equidistant from B-centres, of course, and this places them at the mid-points of the sides of the hexagons which form the market areas of B-centres. This is shown in Figure A2.33.

We can calculate the k-value, as before, either by counting (shares of, where appropriate) dependent settlements or proportions of market areas served with lower order goods. Each K-centre is now served by two B-centres instead of three, which gives each B-centre a half share. There are six of these making three, plus the K-functions of the B-centre itself, giving $k = 4$. The calculation can be carried out for market areas using the information in Figure A2.33.

It is also clearly visible from the figure, using some results of elementary geometry, that if the

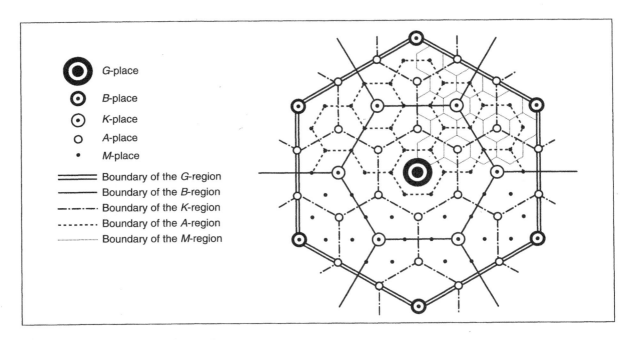

Figure A2.32 A k = 3 system. (From Christaller, 1933)

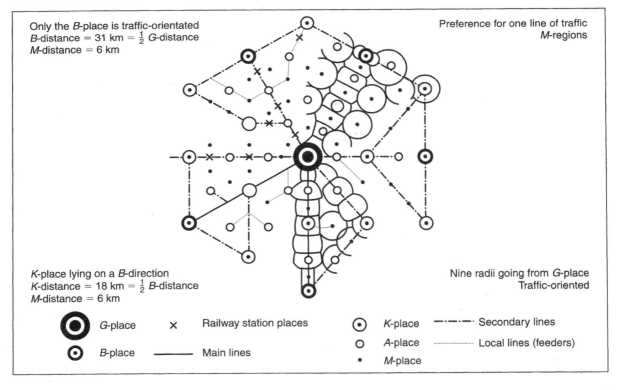

Only the *B*-place is traffic-orientated
B-distance = 31 km = ½ *G*-distance
M-distance = 6 km

Preference for one line of traffic
M-regions

K-place lying on a *B*-direction
K-distance = 18 km = ½ *B*-distance
M-distance = 6 km

Nine radii going from *G*-place
Traffic-oriented

◎	*G*-place	✕	Railway station places	
◉	*B*-place	──	Main lines	

⊙	*K*-place	─·─·─	Secondary lines
○	*A*-place	·········	Local lines (feeders)
•	*M*-place		

Figure A2.33 The transport (k = 4) system. (From Christaller, 1933)

distance between B-centres is *b*, then the distance between K-centres is *b*/2. This factor or ratio can also be usefully noted as √4, since it is, as in the *k* = 3 case, the square root of the *k*-number. Again also, the geometry of the system is such that it can be extended upwards or downwards for any number of steps in the hierarchy.

It is clear from Christaller's own argument that on the whole he thought the marketing principle likely to be more important at least at some levels, and in the corresponding diagram in his book, for example, he shows that the form of organisation can be modified in various ways at lower levels in the hierarchy.

The administration principle: k = 7

Christaller's third alternative form of organisation was based on the notion that a central place might tend to have its dependent places wholly within its market area, through political jurisdiction for example. A system of B-level and K-level centres based on this principle is shown as Figure A2.34. In this case it is easy to see that *k* = 7 by counting dependent settlements and including the B-centre in this. It is slightly more difficult to draw hexagonal market areas within which M-level centres could then be nested.

The factor which relates the distance between centres of different adjacent orders is again √*k*, which is, of course, √7 in this case. Since there is some controversy about how the system is drawn, this should perhaps be interpreted as the factor producing the average distance between lower order centres. (The fact that this sort of relation should hold, even if not exactly, arises from the relation between the distance quantities and the area quantities in this kind of system. Essentially the different tiers are obtained by subdividing market areas at one level into *k* areas at the next lower level. The distance which represents the spacing is then always going to be, on average, at least proportional to, if not equal to, the square root of this number.)

Again it should be noted that, although the system can be tiered through many hierarchical levels, Christaller did not always think this appropriate and in one diagram he shows how a system with a G-centre, with a market area in

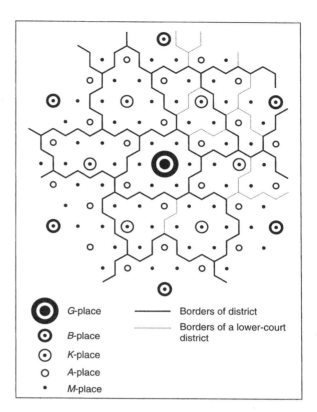

Figure A2.34 The administrative principle: *k* = 7. (From Christaller, 1933)

principle containing six B-centres, can be modified in various ways at the B and lower levels.

General k-systems

This is an appropriate point at which to examine in a more general way how market areas can be built up. This is based on, and in this account anticipates, the work of Losch. It provides both further insight into the basis of Christaller's systems and a basis for the description of Loschian geometry that follows shortly.

We have seen that the settlements in central place systems based on hexagon market area geometry are located at the vertices of a lattice of equilateral triangles (recall Figures A2.27 and A2.33). It is possible to consider in a quite general manner the ways in which hexagons can be drawn to contain various numbers of these settlements. Such a hexagon is called a Dirichlet region and has a degree which turns out to be the k-number we have already met in relation to Christaller's systems. The possible range of such regions is defined subject to the condition that they form a set of areas which cover the points of the lattice in such a way that each region has the same k-number. This is possible for some k-numbers but not for others.

The Loschian system uses k-values generating up to 150 market areas, but we postpone our discussion of this until later as it is first appropriate to add some commentary on the Christaller models.

The Christaller systems: comments and interpretations

The Christaller central place system is essentially determined by too few basic assumptions and cannot hope to be realistic. It offers some valuable insights into the workings of an idealised spatial economy but will inevitably fail to provide an effective model. This is the hypothesis to be explored and tested in the discussion of this section.

Consider first the variety of goods supplied at each level, beginning with the B- and K-levels. Let us label goods by an index ordered according to range, from the least range upwards. Let R_i be the range of the ith good and let T_i be its threshold. Then, at the B-level, goods 1, 2, ... , i(B) will be supplied. The threshold of the $[i(B) + 1]$-good will exceed the maximum real range of level B.

The crucial part of Christaller's analysis is that he starts from level B and, initially at least, works downwards. The spacing of the B-centres (the distance we called b above) will be determined in relation to the range of the good i(B) and its production function – relating revenue and costs at various margins in particular. This calculation is not made explicit (and need not be because it does not affect the principle of the argument), but in effect determines the real range (which is half the spacing in relation to its ideal range and production function). The point to stress is that this calculation is done *for one particular good* (or perhaps groups of goods) *at one particular level*. Thereafter, the whole geometry of the system is determined because the spacing of centres at all other levels is governed by the ratio/factor (assuming the marketing principle). This, in turn,

has followed from the principle that market areas for a particular good are equal in size.

Because the whole geometry is fixed in this way, the mix of goods sold at all other orders of centres is determined (given their ranges) by the initial assumptions as to what constitutes the good i(B) and the spacing of the B-centres. This seems too much to determine by such a pair of assumptions.

Such an argument seems to be the key one and to override other common criticisms of the system in relation to its other assumptions. Many other features of the model could be modified. Beavon (1977), in particular, has shown (albeit for a Loschian system) how varying populations, income levels, transport costs and the like can be introduced and which can modify the initial underlying geometry. In effect, much of this involves laying a Christaller system on a rubber sheet and being prepared to pull and distort it in various ways while still maintaining the same underlying relationships. The main point of criticism spelled out above is then, however, still likely to apply.

THE LOSCHIAN ECONOMIC LANDSCAPE

We saw earlier that Christaller's method is often described as a 'top-down' one. It could be more correctly described as 'middle-down' followed by 'middle-up'. However, Losch's method is undoubtedly 'bottom-up': he begins, not with a population distributed with continuous even density on a uniform plain, but with a set of 'farms' located at the vertices of a lattice of equilateral triangles. We saw how Christaller's decisions about the spacing of B-centres determined the geometry of the whole system; the equivalent decision in Losch's case is therefore the spacing of the farms on the initial lattice. In one sense, this is obviously less restrictive as it does not, of itself, determine the spacing of settlements at a particular level in a hierarchy.

In the previous section we outlined a method for building market areas of successively higher k-values for settlements (or farms) located on a lattice of equilateral triangles. Losch's argument involves calculating successive (in size) market areas which he numbers 1, 2, 3 ... corresponding to $k = 3, 4, 7, ...$ etc., until he has identified 150 such areas (this number is used in the illustration in his book, but it can be assumed there is nothing magical or particularly significant in taking 150 as a limit). He then argues that it is likely that each area will equate with the real range of one or more goods – and if more than one, then they can be classified as the same (i.e. by market area size) for the purpose of the argument. All these market areas can, therefore, in principle occur. The next problem is how to put them together.

A key assumption is that a central metropolis will develop which will be a centre for all goods and so can be used to centre the nets of various sizes. A range of examples of such nets is shown in Figure A2.35. The second step in Losch's argument involves the rotation of some of the nets. The rotations are presumably to be carried out in units of 60° so that a rotated net always has its centres at points on the basic lattice. It is clear from the examples in Figure A2.35 that any nets involving dependent settlements at vertices or edges have orientations which make them invariant with respect to 60° rotations, but that this is not true of other nets. What Losch argues, therefore, is that the nets should be rotated until two different types of 60° sector can be identified: those which have more than an average number of centres, and those with less. The principle of this is shown in Figure A2.35 – in the first six with unrotated nets, and the last three, rotated; an example of the whole picture (a much quoted figure, from Isard's work) is shown in Figure A2.36.

Note that a 'centre' in the Loschian system is very different from Christaller's. A centre is the coincidence of one or more 'sub-centres', let us call them, from the superimposed nets, some of which are being rotated. Thus the goods to be traded at each such centre are nested by enumerating the coinciding sub-centres which make up that particular centre.

Losch appears to carry out this rotation to give more order to his economic landscape than it would otherwise have. He justifies the procedure in part by saying that it minimises transport costs. It is certainly intuitively clear that transport costs are reduced when a greater number of sub-centres are made to coincide by rotation.

A feature of the resulting spatial system is the absence of the neat hierarchical relationships which arise in the Christaller system. There are spatial regularities involving higher order centres in the sense that those consisting of greater numbers of coinciding sub-centres are more widely spaced and appear to form some pattern. However, the neat nesting properties of the Christaller system are absent, and gone is the condition that all higher order centres should sell all lower order goods. It might be argued that the loosening of such restrictions is a good thing: we give a brief assessment of this in the next section.

AN ASSESSMENT OF THE LOSCHIAN SYSTEM

In an earlier section, we criticised the Christaller system for having a geometry that was too rigidly determined by too few assumptions which could be matched against empirical data. The same comment applies here. For all the introduction of a very large number of different market area sizes, because of their mutual relationship in the way the whole landscape is constructed, there are very definite types of centres created at specific locations. There is every reason to expect a variety of centres to exist, and this is an advantage over Christaller, but there is no reason to think it would necessarily take the pattern determined by this geometry. Nor is it then easy to modify the pattern to take account of what might be the reality of a particular situation.

Although the Loschian system was described as 'bottom-up', there is a crucial sense in which it is also 'top-down': this arises from the presumed existence of a central metropolis. It is the location of this, and the set of rotations around it, which give the system its rigidity. It may be more effective to start at another level, at least partially, and carry out some rotations from several centres around the metropolis as well as around the centre itself. This would be difficult to interpret if not to execute, but it might be nearer to reality.

CENTRAL PLACE THEORIES AND EMPIRICAL REALITY

As we have seen, the first formulations of central place theory related to settlements and rural populations. Uniform plain assumptions were, at least in some places, reasonable in relation to agricultural production as the main sector of the

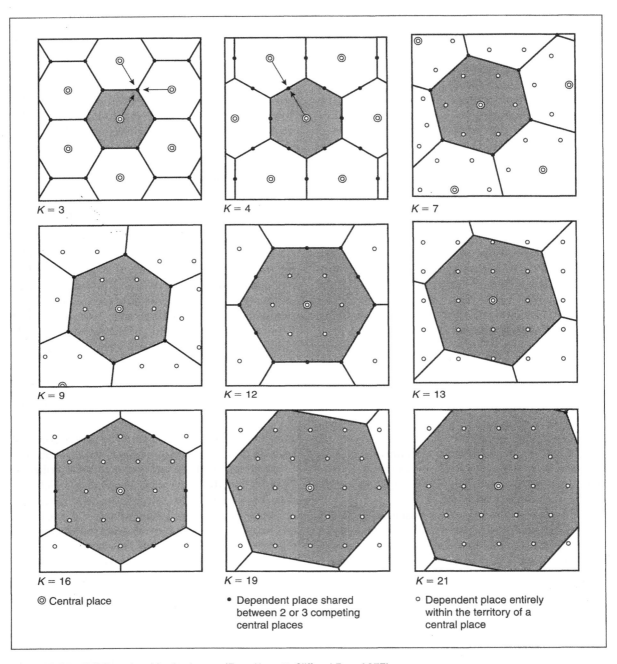

Figure A2.35 Building a Loschian landscape. (From Haggett, Cliff and Frey, 1977)

economy, with settlements providing a marketing and service sector. Now, we would take agriculture to be part of a primary sector which also included resource industries and this alone would distort the uniform plain. The secondary sector would consist of manufacturing industry and there would be important flows of goods between industries for which populations, either rural or in settlements, would not form a suitable proxy in definitions of markets. Then there would be the tertiary sector,

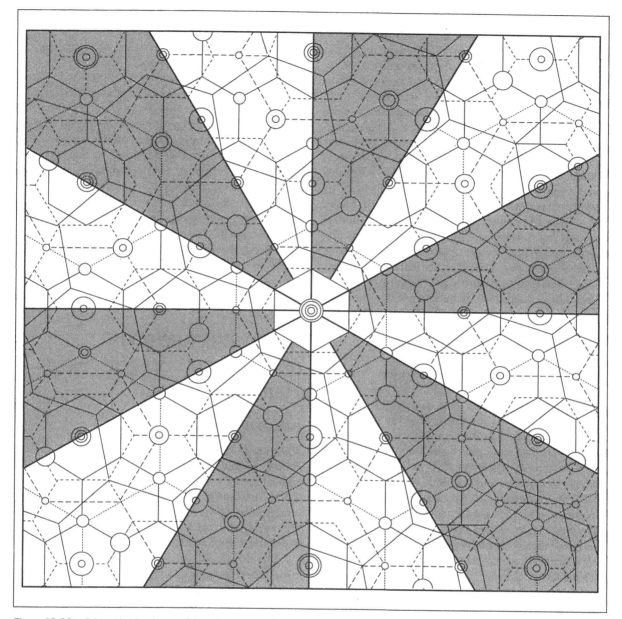

Figure A2.36 A Loschian landscape. (From Isard, 1956)

much of which could be considered to be population-serving, and it is perhaps not surprising that much modern central place theory has been concerned with this sector in urban areas. It may also be necessary to consider explicitly the links between the various sectors, e.g. the role of wholesalers in connecting retailers to their supply of goods. It should also be noted that there has been another more direct kind of modern use of central place theory: in urban planning in some countries it has been used as a basis for designing the location of shopping centres of different orders, and then any underlying 'truth' in the theory becomes a self-fulfilling prophecy.

The empirical investigations associated with central place theory could, in principle, take a number of forms. The most obvious, perhaps, would be to take some form of measurement of settlement sizes and functions and to investigate the applicability of the theories in a direct way. This can be done in relation to pattern, the variety of functions to be found in different centres, and the range of population sizes of centres. This last topic has generated much research into such 'observed' regularities as the rank size rule and controversy (which also relates to mix of functions in centres) as to whether the distributions that have been created are essentially discrete (as Christaller and, to a lesser extent, Losch predict) or continuous. Much of the research is inconclusive on these questions, and given the earlier argument about the rigidity of central place theory, perhaps it could be argued here that these are not the most important questions.

Another route into empirical research would be via the micro-relationships implied by the theory: whether market areas can be related to 'thresholds and ranges' of goods; whether consumers visit the nearest available centre. Again, the answers are much more complicated than the theory predicts. Indeed, it is perhaps surprising that the theory has survived with such a simple sub-theory of market areas given, for example, Palander's (1935) work which charted, theoretically, a much greater variety of shapes that could be generated in different circumstances. And consumers do not all travel to the nearest centre. This will be at least in part due to differences in the quality of centres and the possibilities of multi-purpose trips to higher order centres, which are not directly taken into account in the theory.

SOME CONCLUDING COMMENTS

We argued earlier that Christaller's system is too rigid to have any chance of representing reality. Christaller himself, of course, was well aware of the simplifying assumptions that formed the basis of his theory and gave much attention to relaxations and extensions. It is perhaps surprising that more emphasis has not been given to this. However, it should be emphasised that the theory is an outstanding creation, offering great insights –

rather in the manner of many economic concepts and theories which are accepted on such a basis without ever having a chance of representing reality in any detailed respect.

The Loschian version of the theory is often treated with more respect by textbook writers, but seems to bear even less chance of representing reality and fails to offer the same degree of insight as Christaller's model. Indeed many authors and commentators seem too bemused by the dazzling audacity and brilliance of the formulation to offer much critical comment!

What is needed is a much more flexible basis within which the questions tackled by these early theories can be addressed. It can be argued that there are possible bases that will facilitate this, but a detailed presentation of this argument – which turns on alternative representations of the spatial system – is reserved for chapter 7.

Urban structure and development: Burgess, Hoyt, Harris and Ullman

INTRODUCTION

Burgess, Hoyt, and Harris and Ullman had their major works on the subject of urban form published in 1927, 1939 and 1945 respectively. There was also an earlier and very important contribution by Hurd published in 1903, but on the whole this fits better with a discussion on the development of economic spatial theory.

The three contributions all come from different disciplines: Burgess is a representative of the work of the school of Chicago sociologists; Hoyt was a land economist and real estate man; Harris and Ullman were geographers. Hurd's contribution was based on the notion of rent. In logic, one might have expected a concentric ring theory, based on rent, to follow the work of von Thunen and an agglomeration theory in relation to centres to be based on the work of Weber, but these did not materialise. The Chicago sociologists looked towards plant ecology for their models; Hoyt and Harris and Ullman based their models on detailed empirical observation.

THE CONCENTRIC RING THEORY

The basis of Burgess' spatial theory is the set of ecological concepts used by the Chicago sociologists of the time, the other two main authors being Park and McKenzie. They assumed that the city was populated by distinct groups who were in competition with each other. Particular groups could establish 'dominance' in particular areas so that a distinctive residential pattern was generated. The development process was considered to be driven by upward social mobility, and it was argued that this was achieved through a process of outward migration. The growth of population in cities at the time of writing was such that the vacant spaces left by out-migrants were filled by new incoming migrants. The outward migration process was described in ecological terms as a process of 'invasion and succession' – the 'succession' being the achievement of dominance in the new area by a particular group.

This process, according to Burgess, generated five concentric rings (Figure A2.37). The inner zone was made up of the central business district and was essentially commercial. The second zone was a mixture of industrial and residential activity, and was described also as the 'zone of transition': it was under pressure from the commercial sector as it expanded outwards. The upwardly mobile population left it, and it was the receiving area for new in-migrants to the city. The third zone was made up of working men's homes, established by the first generation of 'successful' workers. The fourth and fifth zones were higher class residential, the outer of the two being known as the 'commuters' ring'.

Burgess and his colleagues gave much attention to particular kinds of areas within each ring. These were often identified by racial or national groupings, or perhaps types of residential stock. The concentric ring pattern of urban land use was seen to be modified by other kinds of patterns. It was often then the role of Burgess and his colleagues to focus on particular small areas of this kind as the basis of their sociological work. In a sense, the concentric ring theory is a by-product.

Perhaps the main point of interest to be taken from Burgess' theory is the emphasis on development: it was a crude *dynamic* theory focusing on the processes of change. The examples of classical theory in earlier sections have been largely static. It is clear that cities do grow outwards and therefore it is not surprising that they have some kind of concentric ring structure. We will see below, however, that there are many features which modify this.

THE SECTOR THEORY

Hoyt's work was based on the study of rents and land values in a large number of American cities. A number of examples, selected from Hoyt's work by Carter (1975), are shown as Figure A2.38. These show the residential distributions of population and display very clear sectoral patterns. A number of reasons can be put forward as to why this is the case. In the early stages of development, the city will be relatively small. Higher class residential development is likely to be attracted by the better physical environments that are available and to be repelled by such land uses as heavy industry. Subsequent development, as the city expands, is likely to follow communication lines such as radial roads and rail routes. In the case of the higher class areas, the only possible direction of expansion is outwards as there will be lower class housing or other land use at either side.

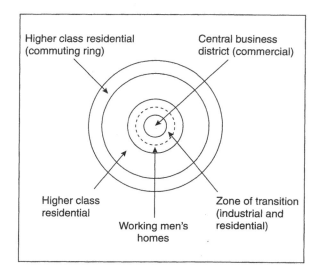

Figure A2.37 Burgess' urban rings

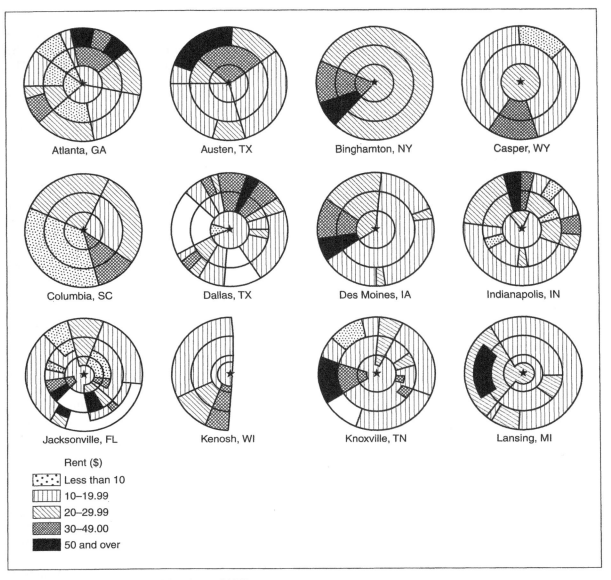

Figure A2.38 Carter's examples (after Carter, 1975)

The dynamic process underpinning Hoyt's model is essentially the same as that of Burgess: that upwardly mobile people are forced to move further out, but the pattern is now a more complicated one. Again as in the Burgess case, it is important to have theories proposed which recognise the influence of sociological variables, like status and class, rather than the pure emphasis on accessibility which is the characteristic of many economic models. Ultimately, our task is to integrate such features of theories.

THE MULTIPLE-NUCLEI THEORY

Harris and Ullman (1945) emphasised another aspect of urban spatial structure: that there are more centres than the original central business district in most large cities. These can arise for at least two kinds of reasons. First, a growing city often absorbs villages and small towns as it expands, and their centres remain. Secondly, centres can develop through processes of agglomeration in the manner of Weber, or through the evolution of a hierarchy, as in central place theory. It is likely, therefore, that if there are concentric ring and sectoral features in urban development, these can be based on multiple nuclei to produce a more complicated overall pattern. Harris and Ullman illustrated their ideas with a much-quoted diagram which is reproduced as Figure A2.39. In that illustration, they emphasise the development of centres of specialisation and the influence of these on urban structure.

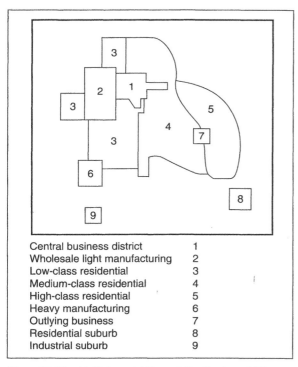

Central business district	1
Wholesale light manufacturing	2
Low-class residential	3
Medium-class residential	4
High-class residential	5
Heavy manufacturing	6
Outlying business	7
Residential suburb	8
Industrial suburb	9

Figure A2.39 Multiple nuclei theory (after Harris and Ullman, 1945)

CONCLUDING COMMENTS

These three theories have been treated relatively briefly because their influence is likely to be of mainly historical interest, though they are still much taught in textbooks in a relatively uncritical way. We can learn a number of things from them.

First, both Burgess and Hoyt at least showed a concern with process and dynamics which was absent from other theories of their period, and also noted the influence of sociological variables as well as economic ones. It is also clear that aspects of all three theories contain elements of truth.

The main difficulty in this kind of approach is the treatment of space. Essentially, the area of the city is being taken as continuous and land-use zones are being identified as rings or sectors, modified by a centre structure. This approach is difficult to carry through dynamically when these areas are shifting in size and location. More importantly, it has been argued that many of the zones, in the way they are defined by the different authors, are difficult to discover empirically, at least in any sharp way. This suggests that the level of resolution is too coarse, and that the form of spatial representation used is inadequate. What we need to look for is better representations and better theories; but in these new theories, the main ideas of Burgess, Hoyt and Harris and Ullman can be broadly reproduced. It would be surprising if this were not the case. In that sense, their contributions are each important, but only as a very rough basis in the search for something better.

Interaction: the gravity model

HISTORY: THE BASIC MODEL

Spatial interaction models have a very long history which, for *gravity* models *per se* culminated in the 1940s and 1950s. This is a little late for classical theory, but it is useful to see the primitive basis on which more productive modelling was built.

There have been two major reviews of the early work on the gravity model, by Carrothers (1956) and by Olsson (1965). Chapter 11 of the book by Isard (1960) is also valuable in this respect. Here, we mainly follow Carrothers. He attributes the earliest statement of the gravity model to Carey in 1858. He quotes Carey as follows:

> The greater the number collected in a given space, the greater is the attractive force which is there exerted … Gravitation is here, as everywhere, in the direct ratio of the mass and the inverse one of distance.

The early applications of the idea were first to migration, by Ravenstein in 1885 and Young in 1924, and to retailing by Reilly in 1929. The two migration models take a similar form. Ravenstein's basic equation can be written (though Carrothers' implies he was not explicit in this):

$$M_{ij} = f(P_i)f(P_j)/d_{ij} \qquad (A2.23)$$

and Young's

$$M_{ij} = kP_iP_j/d_{ij}^2 \qquad (A2.24)$$

where in each case M_{ij} is the number of migrants from city i to city j, P_i is the population of city i, and d_{ij} is the distance between them. In equation (A2.23), $f(P_i)$ represents some function of P_i (and this can be assumed to absorb a constant), and in equation (A2.24), k is a constant. The measures of population, P_i, are being used like a 'mass' in Carey's sense (and he seems to imply in the quotation that there should only be one of them in a formula) though there is no formal reference to a law of gravity.

These simple formulae illustrate most of the features of gravity models. First, you need a measurement of the 'mass' term, and Ravenstein explicitly recognises that this may not be as simple as 'population' on its own. Secondly, you need to decide what the distance function is. This, as a question, is raised by the pair of examples, since one of them shows the migration flow to be inversely proportional to distance, the other inversely proportional to distance squared. Which? A question to which we shall return.

Reilly, interestingly, used gravitational concepts not directly to represent flows but to demarcate retail market areas. Given cities i and j, and populations P_i and P_j respectively, what is the point on a line joining them which is on the boundary of the two market areas? Let this point be d_{ix} from i and d_{xj} from j. Then Reilly's 'law' can be stated as

$$P_x/d_{ix}^2 = P_y/d_{xj}^2 \qquad (A2.25)$$

Note that he uses 'inverse distance squared', and populations as a measure of the attractive power of retailing centres. It could be argued that this equation relates to a typical consumer located at x. In this case, if P_x is the population of x, and we assume that x is on the i-side of the line, the flow might be taken as

$$S_{ix} = kP_iP_x/d_{ix}^2 \qquad (A2.26)$$

where the S_{ix} represents the flow of consumers from i to x and k is a suitable constant.

This now demonstrates a new feature of gravity models: that there are *two* mass terms involved, which makes it more like Newton's original.

The gravity concept was formalised, in these terms, by Stewart and Zipf in the 1940s. Stewart was in fact a physicist who could use the analogy with some direct roots back to his knowledge of physics; Zipf was a sociologist who related social gravity flows to a 'principle of least effort'. This work led to a series of applications to all conceivable kinds of flows: passenger flows, retail flows, telephone calls and newspaper circulation are examples. It had also been applied by Bossard (1932) in an investigation of the function of distance in marriage selection.

POTENTIAL CONCEPTS

Stewart (1942) introduced the concept of potential, based on gravitational laws. Potential in this sense is defined as the 'possibility of interaction' and is defined as:

$$V_{ij} = P_j/d_{ij}^2 \qquad (A2.27)$$

where V_{ij} is the potential of interaction from i (say for one person) to j. It is then possible to calculate a total potential as

$$V = \Sigma_j V_{ij} \qquad (A2.28)$$

Note that, in arguments based on physics, the potential formulae have an inverse distance term, when the flow is based on inverse distance *squared*.

The measure V_i in equation (A2.28) measures *accessibility* from i to population. However, the concept was never really used seriously in this way until after the publication of Hansen's paper in 1959.

DEVELOPMENTS

There are at least two obvious ways in which the traditional gravity model can be developed: first, through an attack on the notion of 'mass'; secondly, by scrutinising the nature of the distance function. In the first case, most people seemed surprisingly wedded to the idea of 'population' as a mass for a very long time. This was broken by Harris in 1964 when he used retail sales in cities as the 'destination' mass term when calculating potentials that gave a measure of market areas. The distance function also had an obvious weakness. There are fundamental reasons in physics why an inverse square law should apply. There are no such reasons in the social sciences. The first step, therefore, is to replace '2' in the formulae we have seen by β, where β is a parameter. The second step along this line is to replace the power function by a more general one, say $f(d_{ij})$, and to investigate possible forms both theoretically and empirically. Finally, in the same way in which we can ask questions about the correct measurement of mass, we can ask corresponding questions about distance. It is possible to use time or travel cost for example. However, these thoughts are substantially overtaken by the presentation of contemporary thinking in Chapter 6.

CONCLUDING COMMENTS

It is interesting that most of the applications of the early gravity models were to inter-city flows. It is as though the authors were happy to deal with spatial systems that were made up of points, but few considered the discrete zoning systems which were to be the foundations of later work. There is one important feature of the spatial representation though. Although gravity models were mainly, at this stage, concerned with flows, they were used relatively early in connection with retail flows, and in this sense they were used to represent market areas. It says something for the modes of thought

at the time that in the first such application by Reilly, the model was used only implicitly as a flow model, with the main job of demarcating non-overlapping market areas. In practice, one of the main achievements of the gravity model approach has been to represent the more realistic *overlapping* market areas.

Perhaps the main surprise in the early applications is that authors restricted themselves for so long to very simple formulae that were dimensionally wrong. In equation (A2.26), for example, if both populations are doubled, then the retail flows are quadrupled, which does not seem intuitively reasonable. The means were available to rectify this by using proportions, but these difficulties were only resolved when the model was taken up in a big way by civil engineers and others working on the transportation studies of the 1950s.

It is also perhaps surprising, especially to this author, that new analogies in physics were not sought earlier. It is interesting that Carrothers, in concluding his review in 1956, wrote:

> ... the gravity and potential concepts of human interaction were developed originally from analogy to Newtonian physics of matter. The behaviour of molecules, individually, is not normally predictable, but in large numbers their behaviour is predictable on the basis of mathematical probability. Similarly, while it may not be possible to describe the actions and reactions of the individual human in mathematical terms, it is quite conceivable that interactions of groups of people may be described in this way.

This sounds more like Boltzmann than Newton – a point to be savoured in relation to the discussion of entropy-maximising in Chapter 6.

Appendix 3

Inter-regional demographic and economic models

We start with population accounts and use this example to illustrate the solutions to certain geographical problems, though the range of application is much wider.

Consider the following simple piece of geography: we wish to examine the variation in birth and death rates in a system of regions $\{i\}$ (or $\{j\}$). Suppose for simplicity, we are dealing with the female population only. If the number of total births in region i is B_i, total deaths D_i, and total population P_i, then we might take

$$b_i = B_i/P_i \qquad (A3.1)$$

as the birth rate in each i, and

$$d_i = D_i/P_i \qquad (A3.2)$$

as the death rate.

However, in a real system, there is migration; and in the period in which B_i and D_i are measured, there will be an array of migration flows, say $\{K_{ij}\}$, where K_{ij} is the number of people resident in i at the beginning of the period and resident in j at the end. (Note that the non-movers are part of this array as K_{ii}.) A little thought shows that this means that the numerators and denominators on the right-hand sides of equations (A3.1) and (A3.2) do not match. Some of the recorded births will be to in-migrants; some of the population will have migrated away and be 'at risk' of giving birth elsewhere. If all the flows are exactly balanced, this may not matter; but typically, it does matter.

The problem can be rectified by using the migration array $\{K_{ij}\}$, but also by extending the accounts to include those who were born and/or who died during the period. (The definition of K_{ij} implies a population which existed at the beginning of the period and survives at the end.) Thus, let $\{K_{\beta(i)j}\}$ be an array such that $K_{\beta(i)j}$ is the set of people who were born in i during the period and survived and were resident in j at the end. Similarly, define $K_{i\delta(j)}$, with $\delta(j)$ representing death recorded in j during the period and $K_{\beta(i)\delta(j)}$ as (typically) infant mortality – those who were born and died in the period. A full set of accounts is then

$$\begin{matrix} \{K^{ij}\} & \{K_{i\delta(j)}\} \\ \{K_{\beta(i)j}\} & \{K_{\beta(i)\delta(j)}\} \end{matrix} \qquad (A3.3)$$

It can immediately be seen that by careful definition of the range of initial and final states, a comprehensive set of accounts, recording what can happen to the basic entities of the system of interest, can be immediately built up. It solves the problem of birth and death rates in that we can now find numerator and denominators which match. For instance

$$b_{ii} = I_{\beta(i)i}/K_{ii} \qquad (A3.4)$$

is the birth rate appropriate to women who were alive in i at the beginning of the period and still resident in i at the end. However, this only leads us to another difficulty: $K_{\beta(i)i}$ is not measured, only B_i. This can be resolved with some difficulty by adding appropriate hypotheses and making some approximations (cf. Rees and Wilson, 1977).

An important general point to establish, however, is that whatever theory is to be developed often needs an accounting basis to make it more effective, and internally consistent. In the demographic case, theories had to be added on birth and death rates – and suitable spatial interaction models need to be developed for the migration flow, $\{K_{ij}\}$.

We now turn briefly to economic accounts. We begin by looking at accounts for a single-region economy with sectors labelled m (or n). Let Z^{mn} be the flow of goods from sector m to sector n – for simplicity, measured in money units. Let Y^m be final demand in sector m and X^m be the total produced. Then the basic accounting equation is

$$\Sigma_n Z^{mn} + Y^m = X^m \tag{A3.5}$$

A multi-regional accounting equation can then be developed by adding subscripts i and j as appropriate:

$$\Sigma_{jn} Z_{ij}^{mn} + \Sigma_j Y_{ij}^m = X_i^m \tag{A3.6}$$

with obvious definitions. When i and j are added, the elements Z_{ij}^{mn} now represent trade flows.

These definitions imply that the product of a sector can be characterised as a 'good' or unified 'bundle' of goods. In some cases, it will be appropriate to take the further step and recognise that each sector has a variety of inputs and a variety of outputs, and to define the basic array element as Z_{ij}^{mng}, the flow from sector m in i to sector n in j of good g.

The task of building models to represent theories is a difficult one, not least because, unlike the demographic case, there is no systematic measurement of inter-regional economic flows at the intra-national scale. However, the development of such theories is clearly an important task and the issue is discussed in Chapter 7. Here, we simply give the broadest idea of what is involved by returning to the aggregate single-region accounts (A3.5). As with the demographic model, the trick is to define suitable rates. Let

$$a^{mn} = Z^{mn}/X^n \tag{A3.7}$$

be the amount of m needed as an input to produce a unit of n. Then we can re-arrange this as follows

$$Z^{mn} = a^{mn}X^n \tag{A3.8}$$

and substitute for Z^{mn} in (A3.7):

$$\Sigma_n a^{mn}X^n + Y^m = X^m \tag{A3.9}$$

These are linear simultaneous equations in $\{X^m\}$ which can be solved. In matrix notation:

$$\mathbf{X} = (\mathbf{I}\text{-}\mathbf{A})^{-1}\mathbf{Y} \tag{A3.10}$$

The model is known at the input–output model and the elements a^{mn} as input–output coefficients. The geographer's task is to extend this economic model to handle trade flows in a system of regions, and considerable progress has been made with this.

The model (A3.10) has been built on a simple idea: first construct the accounts, then define the rates. But it is of considerable power. Suppose there is an increase in final demand in one sector. Production will rise in that sector; additional inputs are pulled in so production has to increase in supplying sectors; those sectors then require additional inputs; and so on. This complicated infinite regress is in fact dealt with by the model (A3.10). This is something which should not be lost – a set of such 'technical' relationships should be present in some form in any theory of a regional economy or of the economy of a system of regions.

As a final part of the preliminary argument, we should also note that the single-region demographic and economic input–output models serve as an important backcloth for an intra-urban model. So there is at least that connection between the two scales. Ultimately, it should be possible to articulate a full connection between the two scales: a set of intra-urban models with aggregates connected to an overall inter-regional model. However the full specification of such a system remains a matter for research.

References

Aleksander, I. and Morton, H. (1990) *An introduction to neural computing*, Chapman and Hall, London.

Alexander, C. (1964) *Notes on the synthesis of form*, Harvard University Press, Cambridge, MA.

Allen, P. M. and Sanglier, M. (1972) Urban evolution, self-organisation and decision-making, *Environment and Planning*, A, 13: 167-83.

Alonso, W. (1960) A theory of the urban land market, *Papers, Regional Science Association*, 6: 149–57.

Alonso, W. (1964) *Location and land use*, Harvard University Press, Cambridge, MA.

Anas, A. (1996) Modeling in urban and regional economics, in Arnott, R. (ed.) *Regional and urban economics*, Harwood Academic, Amsterdam, pp. 921–1050.

Anderson, P. W., Arrow, K. J. and Pines, D. (eds) (1988) *The economy as an evolving complex system*, Addison Wesley, Menlo Park, CA.

Angel, S. and Hyman, G. M. (1970) Urban velocity fields, *Environment and Planning*, 2: 211–24.

Angel, S. and Hyman, G. M. (1972) Urban spatial interaction, *Environment and Planning*, 4: 99–118.

Angel, S. and Hyman, G. M. (1976) *Urban fields*, Pion, London.

Arnott, R. (ed.) (1996a) *Regional and urban economics, Part 1*, Harwood Academic, Amsterdam.

Arnott, R. (ed.) (1996b) *Regional and urban economics, Part 2*, Harwood Academic, Amsterdam.

Arthur, W. B. (1988) Urban systems and historical path dependence, in Ausubel, J. H. and Herman, R. (eds) *Cities and their vital systems: infrastructure, past, present and future*, National Academy Press, Washington, DC.

Arthur, W. B. (1990) Positive feedbacks in the economy, *Scientific American*, February.

Arthur, W. B. (1994a) *Increasing returns and path dependence in the economy*, University of Michigan Press, Ann Arbor, MI.

Arthur, W. B. (1994b) Inductive reasoning and bounded rationality, *American Economic Association Papers and Proceedings*, 84: 406–11.

Arthur, W. B., Ermoliev, Y. M. and Kaniovski, Y. M. (1983) A generalised urn problem and its applications, *Cybernetics*, 19: 61–71.

Arthur, W. B., Durlauf, S. and Lane, D. (1997) Introduction: process and emergence in the economy, in Arthur, W. B., Durlauf, S. and Lane, D. (eds) *The economy as an evolving system II*, Addison Wesley, Reading, MA.

Ashby, W. R. (1956) *Design for a brain*, Chapman and Hall, London.

Axhausen, K. W. and Garling, T. (1992) Activity-based approaches to travel analysis: frameworks, models and research problems, *Transport reviews*, 12: 323–42.

Bak, P. (1997) *How nature works: the science of self-organised criticality*, Oxford University Press, Oxford.

Ball, P. (1999) *The self-made tapestry: pattern formation in nature*, Oxford University Press, Oxford.

Barra, T. de la (1989) *Integrated land use and transport modelling*, Cambridge University Press, Cambridge.

Barrow, J. (1991) *Theories of everything*, Oxford University Press, Oxford.

Batty, M. (1976) *Urban modelling: algorithms, calibrations, predictions*, Cambridge University Press, Cambridge.

Batty, M. (1989) Urban modelling and planning: reflections, retrodictions and prescriptions, in Macmillan, W. B. (ed.) *Remodelling geography*, Blackwell, Oxford, pp. 147–69.

Batty, M. (1994) A chronicle of scientific planning: the Anglo-American modelling experience, *Journal of the American Institute of Planners*, 60: 7–16.

Beavon, K. S. O. (1977) *Central place theory: a reinterpretation*, Longman, London.

Becker, G. S. (1965) A theory of the allocation of time, *Economic Journal*, 75: 488–517.

Beltrami, E. (1987) *Mathematics for dynamic modelling*, Academic Press, Boston.

Berechman, J., Kohno, H., Button, K. J. and Nijkamp, P. (eds) (1996) *Transport and land use*, Edward Elgar, Cheltenham.

Bernstein, R. (1976) *The restructuring of social and political theory*, Blackwell, Oxford.

Berry, B. J. L. (1967) *Geography of market centres and retail distribution*, Prentice Hall, Englewood Cliffs, NJ.

Berry, B. J. L. and Rees, P. H. (1969) The factorial ecology of Calcutta, *American Journal of Sociology*, 74: 445–91.

Bertalanffy, L. von (1968) *General system theory*, Braziller, New York.

Bertuglia, C. S., Leonardi, G., Occelli, S., Rabino, G. A., Tadei, R. and Wilson, A. G. (eds) (1987) *Urban systems: contemporary approaches to modelling*, Croom Helm, London.

Bertuglia, C. S., Leonardi, G. and Wilson, A. G. (1990) *Urban dynamics: designing an integrated model*, Routledge, London.

Bertuglia, C. S., Clarke, G. P. and Wilson, A. G. (eds) (1994) *Modelling the city: performance, policy and planning*, Routledge, London.

Birkin, M. and Wilson, A. G. (1986a) Industrial location models I: a review and an integrating framework, *Environment and Planning, A*, 18: 175–205.

Birkin, M. and Wilson, A. G. (1986b) Industrial location models II: Weber, Palander, Hotelling and extensions in a new framework, *Environment and Planning, A*, 18: 293–306.

Birkin, M. and Wilson, A. G. (1989) Some properties of spatial–structural–economic–dynamic models, presented to the 4th European Colloquium on Quantitative and Theoretical Geography, Veldhoven, 1986, in Hauer, J. *et al.* (eds), *Urban dynamics and spatial choice behaviour*, Kluwer Academic, Dordrecht, pp. 175–201.

Birkin, M., Clarke, M. and Wilson, A. G. (1984) Interacting fields: comprehensive models for the dynamical analysis of urban spatial structure, Working Paper 385, School of Geography, University of Leeds.

Birkin, M., Clarke, G. P., Clarke, M. and Wilson, A. G. (1996) *Intelligent GIS: location decisions and strategic planning*, Geoinformation International, Cambridge.

Bossard, J. H. S. (1932) Residential propinquity as a factor in marriage selection, *American Journal of Sociology*, 38: 219–44.

Boulding, K. (1968) General systems theory – the skeleton of science, in Buckley, W. (ed.), *Modern systems research for the social scientist*, Aldine, Chicago, pp. 3–10.

Boyce, D. E. (1978) Equilibrium solution to combined urban residential location, modal choice and trip assignment models, in Buhr, W. and Freidrich, P. (eds) *Competition among small regions*, Nomos, Baden-Baden, pp. 246–64.

Boyce, D. E. (1984) Urban transportation network-equilibrium and design models: recent achievements and future prospects, *Environment and Planning, A*, 16: 1445–74.

Boyce, D. E., Day, N. D. and MacDonald, C. (1970) *Metropolitan plan making: an analysis of the experience with the preparation and evaluation of alternative land use and transportation plans*, Regional Science Research Institute, Philadelphia.

Brouwer, L. E. J. (1910) Über eineindeutige stetige Transformationen von Flächen in sich, *Mathematische Annalen*, 69: 176–80.

Buck, G. (1998) Most smooth closed-space curves contain approximate solutions to the n-body problem, *Nature*, 395: 51–3.

Buckley, W. (ed.) (1968) *Modern systems research for the social scientist*, Aldine, Chicago.

Burgess, E. W. (1927) The determinants of gradients in the growth of a city, *Publications, American Sociological Society*, 21: 178–84.

Burton, I. (1963) The quantitative revolution and theoretical geography, *Canadian Geographer*, 7: 151–62.

Carey, H. C. (1858) *Principles of social science*, Lippincott, Philadelphia.

Carroll, J. D. (1955) Spatial interaction and the urban-metropolitan regional description, *Papers, Regional Science Association*, 1: 1–14.

Carrothers, G. A. P. (1956) An historical review of the gravity and potential concepts of human interaction, *Journal of the American Institute of Planners*, 22: 94–102.

Carter, H. (1975) *The study of urban geography*, Edward Arnold, London.

Casti, J. L. (1995) *Complexification*, Abacus, London.

Casti, J. L. (1996) *Five golden rules: great theories of twentieth century mathematics*, John Wiley, New York.

Casti, J. L. (1997) *Would-be worlds*, John Wiley, Chichester.

Champernowne, A. F., Williams, H. C. W. L. and Coelho, J. D. (1976) Some comments on urban travel demand analysis, model calibration and the economic evaluation of transport plans, *Journal of Transport Economics and Policy*, 10: 267–85.

Chapman, G. T. (1977) *Human and environmental systems*, Academic Press, London.

Chisholm, M. (1962) *Rural settlement and land use*, Hutchinson, London.

Christaller, W. (1933) *Die centralen Orte in Suddeutschland*, Gustav Fischer, Jena; translated by Baskin, C. W., *Central places in Southern Germany*, Prentice Hall, Englewood Cliffs, NJ.

Clarke, G. P., Langley, R. and Cardwell, W. (1998) Empirical applications of dynamic spatial interaction models, *Computers, Environmental and Urban Systems*, 22: 157–84.

Clarke, G. P. and Wilson, A. G. (1987a) Performance indicators and model-based planning I: the indicator movement and the possibilities for urban planning, *Sistemi Urbani*, 2: 79–123.

Clarke, G. P. and Wilson, A. G. (1987b) Performance indicators and model-based planning II: model-based approaches, *Sistemi Urbani*, 9: 138–65.

Clarke, M. and Wilson, A. G. (1983) Exploring the dynamics of urban housing structure in a 56-parameter residential location and housing model, Working Paper 363, School of Geography, University of Leeds.

Clarke, M. and Wilson, A. G. (1985a) A model-based approach to planning in the National Health Service, *Environment and Planning*, B, 12: 287–302.

Clarke, M. and Wilson, A. G. (1985b) The dynamics of urban spatial structure: the progress of a research programme, *Transactions, Institute of British Geographers*, NS 10: 427–51.

Clarke, M., Keys, P. and Williams, H. C. W. L. (1981) Micro-analysis and simulation of socio-economic systems: progress and prospects, in Wrigley, N. and Bennett, R. J. (eds) *Quantitative geography: a British view*, Routledge and Kegan Paul, London.

Coelho, J. D. and Wilson, A. G. (1976) The optimum location and size of shopping centres, *Regional Studies*, 10: 413–21.

Cohen, J. and Stewart, I. (1994) *The collapse of chaos: discovering simplicity in a complex world*, Viking, London.

Coppock, J. T. and Rhind, D. W. (1991) The history of GIS, in Maguire, D. J., Goodchild, M. F. and Rhind, D. W., *Geographical information systems: principles and applications, Volume 1*, Longman, London, pp. 21–43.

Coveney, P. and Highfield, R. (1995) *Frontiers of complexity: the search for order in a chaotic world*, Faber and Faber, London.

Cox, K. R. (1979) *Location and public problems*, Blackwell, Oxford.

Davies, L. (1987) *Genetic algorithms and simulated annealing*, Pitman, London.

Davies, P. J. and Hersch, R. (1981) *The mathematical experience*, Birkhauser, Boston.

Deco, G. and Obradovic, D. (1996) *An information-theoretic approach to neural computing*, Springer-Verlag, New York and Berlin.

Dendrinos, D. (1997) Cities as spatial chaotic attractors, in Kiel, L. D. and Elliott, E. (eds) *Chaos theory in the social sciences,* University of Michigan Press, Ann Arbor, pp. 237–69.

Department of Health (1996) *The Calman report*, HMSO, London.

Duley, C. and Rees, P. H. (1991) Incorporating migration into simulation models, in Stillwell, J. C. H. and Congdon, P. (1991) *Modelling internal migration*, Belhaven Press, London.

Dunn, E. S. (1954) *The location of agricultural production*, University of Florida Press, Gainsville.

Echenique, M., Flowerdew, A. D. J., Hunt, J. D., Mayo, T. R., Skidmore, I. J. and Simmonds, D. C. (1990) The MEPLAN models of Bilbao, Leeds and Dortmund, *Transport Reviews*, 10: 309–22.

Erlander, S. and Stewart, N. F. (1990) *The gravity model in transportation analysis: theory and extensions*, VSP, Utrecht.

Evans, S. P. (1973) A relationship between the gravity model for trip distribution and the transportation model of linear programming, *Transportation Research*, 7: 39–61.

Farrell, J. G. (1970) *The Singapore grip*, Penguin, Harmondsworth.

Favre, A., Guitton, H., Guitton, J., Lichnerowicz, A. and Wolff, E. (1988, translated by B. E. Schwarzbach, 1995) *Chaos and determinism: turbulence as a paradigm for complex systems converging towards final states*, Johns Hopkins University Press, Baltimore.

Fischer, M. M. and Nijkamp, P. (1993) *Geographic information systems, spatial modelling and policy evaluation*, Springer-Verlag, Berlin.

Fotheringham, A. S. (1983) A new set of spatial interaction models: the theory of competing destinations, *Environment and Planning, A*, 15: 15–36.

Forrester, J. W. (1968) *Principles of systems*, Wright-Allen Press, Canmbridge, MA.

Forrester, J. W. (1969) *Urban dynamics*, MIT Press, Cambridge, MA.

Foster, C. D. and Beesley, M. E. (1963) Estimating the social benefit of constructing an underground railway in London, *Journal of the Royal Statistical Society, A*, 126: 46–92.

Fotheringham, A. S. (1983) A new set of spatial interaction models: the theory of competing destinations, *Environment and Planning, A*, 15: 1121–32.

Fox, M. (1995) Transport planning and the human activity approach, *Journal of Transport Geography*, 3: 106–16.

Fratar, T. J. (1954) Vehicular trip distribution by successive approximation, *Traffic Quarterly*, 8: 53–65.

Frieden, B. R. (1998) *Physics from Fisher information: a unification*, Cambridge University Press, Cambridge.

Gallie, W. B. (1952) *Peirce and pragmatism*, Penguin, Harmondsworth.

Gates, W. (1999) *Business @ the speed of thought*, Penguin, London.

Gibbons, R. (1992) *A primer in game theory*, Harvester-Wheatsheaf, Hemel Hempstead.

Giddens, A. (1979) *Central problems in social theory*, Macmillan, London.

Gilbert, N. (1995) Using computer simulation to study social phenomena, in Lee, R. M. (ed.) *Information technology for the social scientist*, UCL Press, London, pp. 208–20.

Gillen, D. W. (1996) Transport infrastructure and economic development: a review of recent literature, *Logistics and Transportation Review*, 32: 39–62

Golden, R. M. (1996) *Mathematical methods for neural network analysis and design*, MIT Press, Cambridge, MA.

Gordon, G. (1995) The development of organization in an ant colony, *American Scientist*, 83: 50–7.

Gottman, J. (1964) *Megalopolis: the urbanized northeast seaboard of the United States*, MIT Press, Cambridge, MA.

Graham, S. and Marvin, S. (1996) *Telecommunications and the city*, Routledge, London.

Guldmann, J. M. (1998) Competing destinations and intervening opportunities interaction models of inter-city telecommunications flows, *Papers, Regional Science Association*, forthcoming.

Habermas, J. (1974) *Theory and practice*, Heinemann, London.

Hagerstrand, T. (1953) *Innovation diffusion as a spatial process*, translated by Pred, A. (1967) Chicago University Press, Chicago.

Haggett, P. (1965) *Locational analysis in human geography*, Edward Arnold, London

Haggett, P. and Chorley, R. J. (1969) *Network analysis in geography*, Edward Arnold, London.

Haggett, P., Cliff, A. D. and Frey, A. (1977) *Locational analysis in human geography*, 2nd Edition, Edward Arnold, London.

Hall, P. (1982) *Urban and regional planning*, Allen and Unwin, London.

Hall, P. (1988) *Cities of tomorrow*, Blackwell, Oxford.

Hall, P. (1998) *Cities in history*, Blackwell, Oxford.

Hammer, C. and Iklé, F. C. (1957) Intercity telephone and airline traffic related to distance and propensity to interact, Sociometry, 20: 306–16.

Hancock, R. and Sutherland, H. (eds) (1992) *Microsimulation models for public policy analysis: new frontiers*, London School of Economics, London.

Hannon, B. and Ruth, M. (1997) *Modelling dynamic biological systems*, Springer, New York.

Hansen, W. G. (1959) How accessibility shapes land use, *Journal of the American Institute of Planners*, 25: 73–6.

Harding, A. (1990) Dynamic micro-simulation models: problems and prospects, Working paper 48, London School of Economics Welfare State Programme, London.

Harel, D. (1987) *Algorithmics: the spirit of computing*, Addison-Wesley, Reading, MA.

Harris, B. (1962) *Linear programming and the projection of land uses*, Paper 20, Penn-Jersey Transportation Study, Philadelphia.

Harris, B. (1964) A model of locational equilibrium for the retail trade, mimeo, Penn-Jersey Transportation Study, Philadelphia.

Harris, B. (1994) The real issues concerning Lee's 'Requiem', *Journal of the American Institute of Planners,* 60: 31–4.

Harris, B. and Wilson, A. G. (1978) Equilibrium values and dynamics of attractiveness terms in production-constrained spatial-interaction models, *Environment and Planning, A,* 10: 371–88.

Harris, C. D. and Ullman, E. L. (1945) The nature of cities, *Annals, American Academy of Political and Social Sciences*, 242: 7–17.

Harvey, D. L. and Reed, M. (1997) Social science as the sudy of complex systems, in Kiel, L. D. and Elliott, E. (eds), *Chaos theory in the social sciences,* University of Michigan Press, Ann Arbour, pp. 295–323.

Hayles, N. K. (ed.) (1991) *Chaos and order: complex dynamics in literature and science*, University of Chicago Press, Chicago.

Herbert, D. J. and Stevens, B. H. (1960) A model for the distribution of residential activity in an urban area, *Journal of Regional Science*, 2: 21–36.

Hesse, M. (1980) *Revolutions and reconstructions in the philosophy of science*, The Harvester Press, Brighton.

Hillis, W. D. (1987) *The connection machine*, MIT Press, Cambridge, MA.

Hillis, W. D. (1999) *The pattern on the stone*, Weidenfeld and Nicholson, London.

Himanen, V. Nijkamp. P. and Reggiani, A. (eds) (1998) *Neural networks in transport applications*, Ashgate, Aldershot.

Holland, J. H. (1995) *Hidden order: how adaptation builds complexity*, Addison-Wesley, Reading, MA.

Holland, J. H. (1998) *Emergence*, Addison-Wesley, Reading, MA.

Hoover, E. M. (1937) *Location theory and the shoe and leather industries*, Harvard University Press, Cambridge, Ma.

Hoover, E. M. (1967) Some programmed models of industry location, *Land Economics*, 43: 303–11.

Hotelling, H. (1929) Stability in competition, *Economic Journal*, 39: 41–57.

Hoyt, H. (1939) *The structure and growth of residential neighbourhoods in American cities*, Federal Housing Administration, Washington, DC.

Hudson, L. (1972) *The cult of the fact*, Jonathan Cape, London.

Huff, D. L. (1964) Defining and estimating a trading area, *Journal of Marketing*, 28: 34–8.

Hurd, R. M. (1903) *Principles of city land values*, The Record and Guide, New York.

Iklé, F. C. (1954) Sociological relationship of traffic to population and distance, *Traffic Quarterly*, 8: 123–36.

Isard, W. (1956) *Location and the space-economy*, MIT Press, Cambridge, MA.

Isard, W. (1960) *Methods of regional analysis*, MIT Press, Cambridge, MA.

Isard, W., Smith, T. E., Isard, P., Tze Hsiung Tung and Dacey, M. (1969) *General theory: social, political, economic and regional*, MIT Press, Cambridge, MA.

Jensen, H. J. (1998) *Self-organised criticality: emergent complex behaviour in physical and biological systems*, Cambridge University Press, Cambridge.

Jin, Y.-X. and Wilson, A. G. (1993) Generation of integrated multispatial input–output models of cities, *Papers in Regional Science*, 72: 351–67.

Johnson, J. and Picton, H. (1995) *Mechatronics: designing intelligent machines*, Butterworth-Heinemann, Oxford.

Kain, J. F. (1987) Computer simulation models of urban location, in Mills, E. S. (ed.) *Handbook of urban and regional economics, Volume II*, Elsevier, Amsterdam, pp. 847–75.

Kain, J. F. and Apgar, W. C. (1985) *Housing and neighbourhood dynamics: a simulation study*, Harvard University Press, Cambridge, MA.

Kauffman, S. A. (1993) *The origins of order: self-organisation and selection in evolution*, Oxford University Press, Oxford.

Kauffman, S. A. (1995) *At home in the universe: the search for the laws of complexity*, Viking, London.

Kiel, L. D. and Elliott, E. (eds) (1997) *Chaos theory in the social sciences*, University of Michigan Press, Ann Arbor.

King, L. J. and Golledge, R. (1978) *Cities, space and behaviour: the elements of urban geography*, Prentice-Hall, Englewood Cliffs, NJ.

Klosterman, R. E. (1994) An introduction to the literature on large-scale models, *Journal of the American Institute of Planners*, 60: 41–4.

Krugman, P. R. (1993) On the relationship between trade theory and location theory, *Review of International Economics*, 1: 110–22.

Krugman, P. R. (1995) *Development, geography and economic theory*, MIT Press, Cambridge, MA.

Krugman, P. R. (1996) *The self-organizing economy*, Blackwell, Oxford.

Kuhn, T. S. (1962) *The structure of scientific revolutions*, University of Chicago Press, Chicago.

Lakshmanan, T. R. and Hansen, W. G. (1965) A retail market potential model, *Journal of the American Institute of Planners*, 31: 134–43.

Lee, D. B. (1973) Requiem for large-scale models, *Journal of the American Institute of Planners*, 39: 163–78.

Lefeber, L. (1958) *Allocation in space*, North Holland, Amsterdam.

Leontief, W. (1967) *Input–output analysis*, Oxford University Press, Oxford.

Leslie, P. H. (1945) On the use of matrices in certain population mathematics, *Biometrika*, 23: 183–212.

Leslie, P. H. (1948) Some further notes on the use of matrices in population mathematics, *Biometrika*, 35: 213–45.

Leung, Y. (1997) *Intelligent spatial decision support systems*, Springer-Verlag, Berlin.

Levy, S. (1992) *Artificial life*, Penguin, Harmondsworth.

Lill, E. (1891) *Das Reisegesetz und seine Anwendung auf den Eisenbahnverkehr*, Vienna; cited in Erlander and Stewart (1990).

Longley, P. and Batty, M. (eds) (1996) *Spatial analysis: modelling in a GIS environment*, Geoinformation International, Cambridge.

Losch, A. (1940) *Die raumliche Ordnung der Wirtschaft*, Gustav Fischer, Jena; translated by Woglam, W. H. and Stolper, W. F. (1954) *The economics of location*, Yale University Press, New Haven, CT.

Lowry, I. S. (1964) *A model of metropolis*, RM-4035-RC, The Rand Corporation, Santa Monica.

Lowry, I. S. (1967) *Seven models of urban development: a structural comparison*, The Rand Corporation, Santa Monica.

Macgill, S. M. and Wilson, A. G. (1979) Equivalences and similarities between some alternative urban and regional models, *Sistemi Urbani*, 1: 9–40.

Mackett, R. L. (1980) The relationship between transport and the viability of central and inner urban areas, *Journal of Transport Economics and Policy*, **14**: 267–94.

Mackett, R. L. (1990) Comparative analysis of modelling land use–transport interaction at the micro and macro levels, *Environment and Planning, A*, **22**: 459–75.

Mackett, R. L. (1993) Structure of linkages between transport and land use, *Transportation Research, B*, **27**: 189–206.

Macmillan, W. D. (1996) Fun and games: serious toys for city modelling in a GTS environment, in Longley, P. and Batty, M. (eds) *Spatial analysis: modelling in a GIS environment*, Geoinformation International, Cambridge, pp. 153–65.

Maguire, D. J., Goodchild, M. F. and Rhind, D. W. (1991) *Geographical information systems: principles and applications, Volume 1*, Longman, London.

Malthus, T. R. (1798) *An essay on the principle of population*, J. Johnson, London.

Martin, L. and March, L. (eds) (1972) *Urban space and structures*, Cambridge University Press, Cambridge.

May, R. M. (1971) Stability in multi-species community models, *Mathematical Biosciences*, **12**: 59–79.

May, R. M. (1973) *Stability and complexity in model ecosystems*, Princeton University Press, Princeton, NJ.

Mills, E. S. (ed.) (1987) *Handbook of urban and regional economics, Volume 2: urban economics*, North Holland, Amsterdam.

Nakamura, H., Hayashi, Y. and Miyamoto, K. (1983) Land use transportation analysis system for a metropolitan area, *Transportation Research Record*, **931**: 12–19.

Nash, J. (1950) Equilibrium points in n-person games, *Proceedings of the National Academy of Sciences*, **36**: 48–9.

Neumann, J. von and Morgenstern, O. (1947) *Theory of games and economic behaviour*, Princeton University Press, Princeton, NJ.

Nijkamp, P. (ed.) (1986) *Handbook of regional and urban economics, Volume 1: regional economics*, North Holland, Amsterdam.

Olsson, G. (1965) *Distance and human interaction: a review and bibliography*, Regional Science Research Institute, Philadelphia.

Openshaw, S. (1988) Building an automated modelling system to explore a universe of spatial interaction models, *Geographical Analysis*, **20**: 31–6.

Openshaw, S. (1992) Some suggestions concerning the development of artificial intelligence tools for spatial modelling and analysis in GIS, *Annals of Regional Science*, **26**: 35–51.

Openshaw, S. (1993) Modelling spatial interaction using a neural net, in Fischer, M. M. and Nijkamp, P. (eds) *Geographic information systems, spatial modelling and policy evaluation*, Springer-Verlag, Berlin, pp 147–66.

Orcutt, G. H. (1957) A new type of socio-economic system, *Review of Economic Statistics*, **58**: 773–97.

Orcutt, G. H., Watts, H. W. and Edwards, J. B. (1968) Data aggregation and information loss, *American Economic Review*, **58**: 773–87.

Osborne, A. D. (1999) *Complex variables and their applications*, Addison Wesley Longman, Harlow.

Paelink, J. H. P. and Nijkamp, P. (1975) *Operational theory and method in regional economics*, Saxon House, Farnborough.

Page, S. E. (1998) On the emergence of cities, mimeo, University of Iowa.

Palander, T. (1935) *Beiträge zur Standortstheorie*, Almquist and Wiksell, Uppsala.

Paulley, N. J. and Webster, F. V. (1991) Overview of an international study to compare models and evaluate land use and transport policies, *Transport Reviews*, **11**: 197–222.

Pooler, J. (1994) An extended family of spatial interaction models, *Progress in Human Geography*, **18**: 17–39.

Porush, D. (1991) Fictions as dissipative structures: Prigogine's theory and postmodernism's Roadshow, in Hayles, N. K. (ed.) *Chaos and order: complex dynamics in literature and science*, University of Chicago Press, Chicago, pp. 54–84.

Poston, T. and Wilson, A. G. (1977) Facility size versus distance travelled: urban services and the fold catastrophe, *Environment and Planning, A,* **9**: 681–6.

Prigogine, I. (1980) *From being to becoming: time and complexity in physical science,* Freeman, San Francisco.

Prigogine, I. and Stengers, I. (1984) *Order out of chaos: man's new dialogue with nature,* Heinemann, London.

Putman, S. H. (1983) *Integrated urban models: policy analysis of transportation and land use,* Pion, London.

Putman, S. H. (1991) *Integrated urban models 2. New research and applications of optimization and dynamics,* Pion, London.

Quine, W. V. (1960) *Word and object,* MIT Press, Cambridge, MA.

Ratcliffe, P. (ed.) (1996) *Ethnicity in the 1991 Census,* HMSO, London.

Ravenstein, E. G. (1885) The laws of migration, *Journal of the Royal Statistical Society,* 48, 167–227.

Rees, P. H. (1979) *Residential patterns in American cities,* RP–189, Department of Geography, University of Chicago.

Rees, P. H. (1996) Projecting the national and regional populations of the European Union using migration information, in Rees, P. H., Stillwell, J. C. H., Convey, A. and Hupiszewski, M. (eds.), *Population migration in the European Union,* John Wiley, Chichester, pp. 331–64.

Rees, P. H. and Phillips, D. (1996) Geographical spread: the national picture, in Ratcliffe, P. (ed.), *Ethnicity in the 1991 census,* HMSO, London.

Rees, P. H. and Wilson, A. G. (1977) *Spatial population analysis,* Edward Arnold, London.

Reilly, W. J. (1929) *Methods for the study of retail relationships,* Bulletin No. 2944, University of Texas.

Reilly, W. J. (1931) *The law of retail gravitation,* G. P. Putnam, New York.

Richardson, H. W. (1977) *The new urban economics and alternatives,* Pion, London.

Rihll, T. E. and Wilson, A. G. (1987a) Spatial interaction and structural models in historical analysis: some possibilities and an example, *Histoire et Mesure,* II–1: 5–32.

Rihll, T. E. and Wilson, A. G. (1987b) Model-based approaches to the analysis of regional settlement structures: the case of ancient Greece, in Denley, P. and Hopkin, D. (eds) *History and computing,* Manchester University Press, Manchester, pp. 10–20.

Rihll, T. E. and Wilson, A. G. (1991) Settlement structures in Ancient Greece: new approaches to the polis, in Rich, J. and Wallace-Hadrill, A. (eds) *City and country in the ancient world,* Croon Helm, London, pp. 58–95.

Robinson, G. M. (1998) *Methods and techniques in human geography,* John Wiley, Chichester.

Rogers, A. (1973) The mathematics of multi-regional demographic growth, *Environment and Planning,* 5: 3–29.

Rogers, A. (1975) *Introduction to multi-regional mathematical demography,* John Wiley, New York.

Ruth, M. and Hannon, B. (1997) *Modelling dynamic economic systems,* Springer-Verlag, New York.

Scarf, H. (1973a) *The computation of economic equilibria,* Yale University Press, New Haven.

Scarf, H. (1973b) Fixed-point theorems and economic analysis, *American Scientist,* 71: 289–96.

Schlager, K. J. (1965) A land use plan design model, *Journal of the American Institute of Planners,* 31: 103–11.

Scott, A. J. (1971) *Combinatorial programming, spatial analysis and planning,* Methuen, London.

Senior, M. L. and Wilson, A. G. (1974) Explorations and syntheses of linear programming and spatial interaction models of residential location, *Geographical Analysis,* 6: 209–38.

Shannon, C. and Weaver, W. (1949) *The mathematical theory of communication,* University of Illinois Press, Urbana.

Smelser, N. J. (1963) *The theory of collective behaviour,* Free Press of Glencoe, New York.

Smith, T. E. and Shang-Hsing Hsieh (1997) Gravity-type interactive Markov models – Part I: a programming formulation for steady states, *Journal of Regional Science,* 37: 683–708.

Stevens, B. H. (1968) Location theory and programming models: the von Thunen case, *Papers, Regional Science Association,* 21: 19–34.

Stewart, J. Q. (1942) A measure of the influence of population at a distance, *Sociometry*, 5: 63–71.

Stone, R. (1967) *Mathematics in the social sciences*, Chapman and Hall, London.

Stone, R. (1970) *Mathematical models of the economy*, Chapman and Hall, London.

Stouffer, S. A. (1940) Intervening opportunities: a theory relating mobility and distance, *American Sociological Review*, 5: 845–67.

Thom, R. (1975) *Structural stability and morphogenesis*, W. A. Benjamin, Reading, MA.

Thomsen, E. (1997) *OLAP solutions: building multidimensional information systems*, John Wiley, New York.

Thunen, J. H. von (1826) *Der isolierte Staat in Beziehung auf Landwirtschaft und Nationalökonomie*, Gustav Fisher, Stuttgart; translation by C. M. Wartenburg (1966) *The isolated state*, Oxford University Press, Oxford.

Timms, D. W. G. (1971) *The urban mosaic*, Cambridge University Press, Cambridge.

Thompson, D. (1942) *On growth and form*, Cambridge University Press, Cambridge.

Varela, F. (1979) *Principles of biological autonomy*, North Holland, Amsterdam.

Volterra, V. (1938) Population growth, equilibria and extinction under specified breeding conditions: a development and extension of the theory of the logistic curve, *Human Biology*, 10: 1–11.

Weaver, W. (1948) Science and complexity, *American Scientist*, 36: 536–44.

Weaver, W. (1958) A quarter century in the natural sciences, *Annual Report, The Rockefeller Foundation*, New York, pp. 7–122.

Weber, A. (1909) *Uber den Standort der Industrien*, Tubingen; translation by Friedrich, C. J. (1929) *Theory of the location of industries*, University of Chicago Press, Chicago.

Webster, F. V., Bly, P. H. and Paulley, N. J. (eds) (1988) *Urban land use and transport interaction*, Gower, Aldershot.

Wegener, M. (1986a) Transport network equilibrium and regional deconcentration, *Environment and Planning*, A, 18: 437–56.

Wegener, M. (1986b) Integrated forecasting models of urban and regional systems, *London Papers in Regional Science*, 15, *Integrated analysis of regional systems*, pp. 9–24.

Wegener, M. (1994) Operational urban models: the state of the art, *Journal of the American Institute of Planners*, 60: 17–29.

Weinberg, S. (1994) *Dreams of a final theory*, Viking, London.

Williams, G. P. (1997) *Chaos theory tamed*, Taylor and Francis, London.

Williams, H. C. W. L. (1977) On the formation of travel demand models and economic evaluation measures of user benefit, *Environment and Planning*, A, 9: 285–344.

Williams, H. C. W. L., Kim, K. S. and Martin, D. (1990) Location–spatial interaction models: 1, 2 and 3, *Environment and Planning*, A, 22: 1079–89, 1155–68 and 1281–90.

Wilson, A. G. (1967) A statistical theory of spatial distribution models, *Transportation Research*, 1: 253–69.

Wilson, A. G. (1970) *Entropy in urban and regional modelling*, Pion, London.

Wilson, A. G. (1971a) A family of spatial interaction models and associated developments, *Environment and Planning*, 3: 1–32.

Wilson, A. G. (1971b) Generalizing the Lowry model, *London Papers in Regional Science*, 2: 121–34.

Wilson, A. G. (1972) Some recent developments in microeconomic approaches to modelling households behaviour with special to spatiotemporal organisation, presented to the Centre for Environmental Studies Conference on Urban Economics, Keele, 1971; reprinted in Wilson, A. G. *Papers in urban and regional analysis*, pp. 216–36.

Wilson, A. G. (1974) *Urban and regional models in geography and planning*, John Wiley, Chichester and New York.

Wilson, A. G. (1975) Learning and control mechanisms for urban modelling, in Cripps, E. L. (ed.) *Regional science: new concepts and old problems*, Pion, London, pp. 137–55.

Wilson, A. G. (1978) Spatial interaction and settlement structure: towards an explicit central place theory, in Karlqvist, A., Lundqvist, L., Snickars, F. and Weibull, J. W. (eds) *Spatial interaction theory and planning models*, North Holland, Amsterdam, pp. 137–56.

Wilson, A. G. (1981a) *Geography and the environment: systems analytical approaches*, John Wiley, Chichester.

Wilson, A. G. (1981b) *Catastrophe theory and bifurcation: applications to urban and regional systems*, Croom Helm, London; University of California Press, Berkeley, CA.

Wilson, A. G. (1983a) A generalised and unified approach to the modelling of service-supply structures, Working Paper 352, School of Geography, University of Leeds.

Wilson, A. G. (1983b) Transport and the evolution of urban spatial structure, in *Atti delle Giornate di Lavoro 1983*, Guida Editori, Naples, pp. 17–27.

Wilson, A. G. (1985a) Structural dynamics and spatial analysis: from equilibrium balancing models to extended economic models for both perfect and imperfect markets, Working Paper 431, School of Geography, University of Leeds.

Wilson, A. G. (1985b) Location theory: a unified approach, Working Paper 355, School of Geography, University of Leeds.

Wilson, A. G. (1985c) Raising the levels of ambition in research: some lessons from the journal, *Environment and Planning, A,* 17: 465–70.

Wilson, A. G. (1988) Configurational analysis and urban and regional theory, *Sistemi Urbani,* 10: 51–62.

Wilson, A. G. (1989) Classics, modelling and critical theory: human geography as structured pluralism, in Macmillan, W. B. (ed.) *Remodelling geography*, Blackwell, Oxford, pp. 61–9.

Wilson, A. G. (1995) Simplicity, complexity and generality: dreams of a final theory in locational analysis, in Cliff, A. D., Gould, P. R., Hoare, A. G. and Thrift, N. J. (eds) *Diffusing geography: essays for Peter Haggett*, Blackwell, Oxford, pp. 342–52.

Wilson, A. G. (1998) Land use-transport interaction models: past and future, *Journal of Transport Economics and Policy,* 32: 3–26.

Wilson, A. G. (1999) Urban modelling: conceptual, mathematical and computational challenges, paper presented to the European Meeting of the Regional Science Association, Dublin, August 1999.

Wilson, A. G. and Bennett, R. J. (1985) *Mathematical methods in human geography and planning*, John Wiley, Chichester and New York.

Wilson, A. G. and Birkin, M. (1987) Dynamic models of agricultural location in a spatial interaction context, *Geographical Analysis,* 19: 31–56.

Wilson, A. G. and Kirkby, M. J. (1975; 2nd Edition, 1980) *Mathematics for geographers and planners*, Oxford University Press, Oxford.

Wilson, A. G. and Macgill, S. M. (1979) A systems analytical framework for comprehensive urban and regional modelling, *Geographica Polonica,* 42: 9–25.

Wilson, A. G. and Oulton, M. J. (1983) The corner-shop to supermarket transition in retailing: the beginnings of empirical evidence, *Environment and Planning, A,* 15: 265–74.

Wilson, A. G. and Pownall, C. M. (1976) A new representation of the urban system for modelling and for the study of micro-level interdependence, *Area,* 8: 256–64.

Wilson, A. G. and Senior, M. L. (1974) Some relationships between entropy maximizing models, mathematical programming models and their duals, *Journal of Regional Science,* 14: 207–15.

Wilson, A. G., Coelho, J. D., Macgill, S. M. and Williams, H. C. W. L. (1981) *Optimization in locational and transport analysis*, John Wiley, Chichester and New York.

Wilson, N. L. (1959) Substances without substrata, *Review of Metaphysics,* 12: 521–39.

Young, E. C. (1924) The movement of farm population, *Bulletin 426*, Cornell University, Agriculture Experiment Station, Ithaca, NY.

Zipf, G. K. (1946) The P_1P_2/D hypothesis on the inter-city movement of persons, *American Sociological Review,* 11: 677–86.

Index

sociology 36, 55
South America 95
spatial competition 53
spatial distribution of activities 20
spatial dynamics of prices and rents 77
spatial flows 17
spatial interaction 21
spatial interaction models 62
    doubly-constrained 62
    early 50
    elementary 56
spatial interaction variables 116
spatial representations 14, 15, 108, 132
spatial scales 3
spatial science 6
spatial structures 73
spatial systems 6
    alternative 16
sport 106
stability under competition 134
standard industrial classification 8
statistical analysis 25
statistical models 98
statistical mechanics 2, 63
statistics 41
steady states, stable 110
Stengers, I. 44, 161
Stevens, B.H. 60, 80, 90, 121, 122, 158, 161
Stewart, I. 108, 156
Stewart, J.Q. 150, 162
Stewart, N.F. 56, 157
Stirling's approximation 63
STM dimensions 14
STM framework 30
STM interaction 58
STM principles, application of 96
Stone, R. 62, 161
Stouffer, S.A. 105, 162
structural dynamics 70
structuralist approach 4, 35
structure 59
substitution principle 34
subjective scoring systems 100
succession 55, 147
sum of squares 109
super-computing power, availability of 98
superhospitals 96
supermarkets 96
supradisciplinary concept 43
surplus value 33
Sutherland, H. 97, 158

system articulation 14, 30
systematic geography 35
systems 2
systems analysis 42, 43
    under-developed 101
systems dynamic school 42, 112
systems' modelling focus 4
systems of interest 4, 6

tagging 46
technical interests 40
technological change 33
telecommunications 105
telecommunications traffic 44
temporal scale 3
theory 14, 37
    building 30, 42
    development 4
    of knowledge and meaning 37
    of the location of industries 52
    in urban and regional analysis 20
theory-laden 40
Thom, R. 41, 162
Thompson, D. 45, 162
Thomsen, E. 97, 162
threshold of a good 136
time consumption 108
time geography 108
time, value of 59
Timms, D.W.G. 60, 162
town planning 48
trade flows in a system of regions 153
traffic 37
transport 91, 108
    costs 54, 81
    flow matrix 62
    as a function of land use 62
    infrastructure 92
    model, disaggregation of 91
    planning 60
    principle 139
    services 34
    studies 48
    system 13, 140
transportation problem of linear programming 65
travel cost 59
travel cost impedance function 63
Treasury manual on project appraisal 99
trip distribution 66, 91
trip generation 66, 91
truth 19, 38, 39

Printed and bound by CPI Group (UK) Ltd, Croydon, CR0 4YY
01/05/2025
01858568-0001